# The Principles and Techniques of
# Engineering Estimating

**Other Titles of Interest**

COWAN
Quality Control for Management

DONALD
Management Information and Systems

GRAHAM
Work Measurement and Cost Control

GOLD
Analyzing Technological Change;    Economics.
Management & Environment

HUGHES
Human Relations in Management

HUNT
Industrial Economics

HUSSEY
Introducing Corporate Planning

KAY
A Mathematical Model for Handling in a warehouse

KING
PRODUCTION    Planning and Control An Introduction to
Quantitative Methods

LUMSDEN
The Line-of-Balance Method

MELLYK
Principles of Applied Statistics

TOWNSEND
Scale, Innovation, Merger and Monopoly

# The Principles and Techniques of Engineering Estimating

*by*

GRANVILLE CALDER, D.M.S., M.A.Cost E., Cert. of Ed.

PERGAMON PRESS

OXFORD · NEW YORK · TORONTO

SYDNEY · PARIS · FRANKFURT

| U.K. | Pergamon Press Ltd., Headington Hill Hall, Oxford OX3 0BW, England |
| U.S.A. | Pergamon Press Inc., Maxwell House, Fairview Park, Elmsford, New York 10523, U.S.A. |
| CANADA | Pergamon of Canada Ltd., P.O. Box 9600, Don Mills M3C 2T9, Ontario, Canada |
| AUSTRALIA | Pergamon Press (Aust.) Pty. Ltd., 19a Boundary Street, Rushcutters Bay, N.S.W. 2011, Australia |
| FRANCE | Pergamon Press SARL, 24 rue des Ecoles, 75240 Paris, Cedex 05, France |
| WEST GERMANY | Pergamon Press GmbH, 6242 Kronberg-Taunus, Pferdstrasse 1, Frankfurt-am-Main, West Germany |

First edition 1976

**Library of Congress Cataloging in Publication Data**

Calder, Granville.
The principles and techniques of engineering estimating.

1. Engineering—Estimates and costs. I. Title.
TA183.C34  1976        620        75-40108
ISBN 0-08-019704-3
ISBN 0-08-019703-5 pbk.

*Printed in Great Britain by* A. Wheaton & Co, Exeter

# CONTENTS

# PREFACE

In this book I have attempted to provide a brief yet comprehensive explanation of the procedures relating to the field of engineering estimating.

Within the narrow limits of space I have allowed myself it has not been possible to go into the minute detail of each and every aspect of the estimating function; these however will be known by the individual practising estimator.

The aim and scope of the book has been:

1. To establish a base upon which estimating terms and definitions can be standardised, so that the estimating function can be treated by management as a true discipline.
2. To bring into the open the ramifications and the total depth of the field of engineering estimating, emphasising the wide sphere of the work involved so that an estimator can have his work appreciated and understood.
3. An attempt is made to formalise engineering estimating into a systematic procedure of events which can be used by any student or participant in the engineering estimating field; which if followed would lead to the implementation of accurate estimates.

Any decision taken at Board level or below which involves finance is normally taken on the basis of evidence placed before them in the form of estimates, it is essential therefore that estimating is a true discipline for management. Because of the manner in which estimating is covered and presented this book should be very useful to any manager, student or estimator.

I would like to offer my thanks to all who have assisted and encouraged me to get this book into print; I should also like to give acknowledgement to the Association of Cost Engineers, Mr Stanley Kirkham, the Modulex Company Ltd., and Bronx Engineering, Lye, Ltd., for allowing me to reproduce their charts and formulae.

G. CALDER

Stourbridge,
West Midlands.

# INTRODUCTION

Within the modern concept of management functions, an excellent place for development is in the estimating department, where one can develop both responsibility and the ability to work under continuing pressures. The estimator comes into contact with every phase of production, work study and all ancillary activities within a company. He is also extremely cost conscious and in close touch with all financial aspects of a company. As such, everyone within a company is extremely interested in the activities of the estimating department, whether from the manufacturing, sales or financial departments. If one is merely a student or an employee working in any department of a company, here is an activity about which one should be more conversant, a field that requires an imaginative, alert and probing analytical mind.

## Historical Development

One of the first items that must have come into the mind of man in his development from a cave dweller to the complex social being of today is the old, old question of "How much is it going to effect me?", whether it is in time, money or material. In all cases, whether the effect is in time, money or material, estimates have to be made. The very first tribal estimates were done purely by custom; it was the custom to hunt for this and that, to store for the long winter in order to feed and clothe the tribe. Custom decreed that the requirements for the tribe was of this amount; was this amount arrived at purely by chance? Trial and error? Or did the old tribal chiefs and elders actually plan and estimate for their requirements?

Later, however, as civilisations evolved, building their cities, where duties, taxes and capital were required in order to advance the civilisation, it is evident that planning and more formalised estimates were used. Estimating, therefore, has been followed throughout the ages until it has become the art that it is today and in my opinion worthy of anyone's attention and dedication.

# INTRODUCTION

## Plan of the Book

This book is an endeavour to formalise the concepts and analyse all the various implications relating to estimating. It follows through all forms of estimating; the principles behind the various forms and the elements that effect them. A picture is drawn to show how estimating fits into the overall management organisation and structure, with all the ramifications that it possesses from the small component estimate to its place amongst technological forecasting. Estimating is a true management technique, upon which rests the whole future and success of the company. Only by a full understanding of the technique can management be successful.

# CHAPTER 1

# MEANING AND SCOPE OF ESTIMATING

Estimating can best be understood by being able to understand the implications and the interactions that take place between the estimating department and all the other departments within a company. It is also important to understand how the outcome of estimates can affect a company, and why and how estimates are made. Upon whether this is clearly understood depends the growth of the estimating function; the accuracy required of it by management will also determine its size. We can therefore learn much about a company by being able to study its estimating department's function.

## The Problem of Defining Objectives

The main objective of any company, if one is asked the question, "is to make the greatest possible profit". To most, this would appear to be true, in order to give the most profitable return to the shareholders' investment.

Peter Drucker in *The Practice of Management*, however, summarises the concepts as:

1. It is the first duty of a business to survive.
2. The guiding principle of business economics is *not* the maximisation of profits; it is the avoidance of a loss.

These concepts are at variance with the classical economic approach, i.e. that a company has to make a profit. A company has to "avoid a loss" in order to survive; it has to survive in order to continue in business.

Therefore profit is of a secondary nature and motive, although highly desirable to all, for economic growth and a return on capital invested. At times a profit motive can lead to a state of stagnation, particularly if estimates are drawn up purely with the idea of making the best possible profit without any relationship to the state of the market. The modern concept of rationalising a company in order to pay a profit on a short-run

1

basis is, in itself, a quick way to the destruction of a company on a long-term venture; as any severity in the rationalisation will lead to the inability of the company to meet any extended expansion in the future, in order to meet competition. The objectives of the company must be laid down clearly so that one can carry out the secondary departmental function. The estimating department, as such, is tied up to both long- and short-range strategies of the company. Because of this it is important that the company's objectives are clearly known to them.

## What Does Management Expect from an Estimating Department?

Management's attitude towards the estimating department has changed drastically over the past few years as the costs of labour and material in production rise and become more and more competitive and intensive. In the past, management has, upon occasions, been able to rely upon intimate knowledge of the trade to be able to quote a price for an article without any formal estimating procedure, particularly where the labour/material content has been negligible.

Good estimating today is a must for economic growth. Errors that are made are costly, both in lost sales, loss of profits or contribution to profits.

Faulty estimating can only be blamed upon bad management, as the estimating function can only be as good as the objectives and policies laid down by management. If the estimating department does not have clear-cut policies and procedures to follow, with the added authority to carry them out, then the result that will be obtained from the estimate will only be of the accuracy that management has decreed and can expect, because of the limitations placed upon the estimating function. Whilst in general, particularly within the production department, control grows and grows, the controls within the estimating department stagnate, as little more than a cursory glance is paid to the estimating function, its education or training. This is because of the unfortunate fact that the estimating department is rarely allowed to, or is unable to, present any figures that can show that the estimating department has, in fact, been responsible for making a profit.

The estimating function is a hypothetical function. It assesses the future by relying upon experience and past history; as such it is not able to defend itself at the time, unless the estimating function is correctly carried out.

It is useless for an estimating department to prepare an estimate, be quoted upon the estimate, if, when the order is obtained, there is no control upon subsequent events and no notice is taken whatsoever of the estimate. Although the estimate should never be taken as a production plan in itself, it should become a cost control document, where the overall costs of labour, overheads and material are kept within the overall estimated limits. If this should turn out to be impossible, there is something wrong, either with the estimating function or the cost controls. It is more normal and easier for the estimating function to be blamed at first, and unless the investigation into why the error has occurred is carried out zealously, it will be covered up before any truthful picture can emerge. This should not be allowed to happen, because if it does, it will proliferate upon the estimating function and in the future the mistakes will become larger and larger. For management the estimating function does not end with the mere production of the estimate. It continues and follows on after the estimate, from the time the quoted price is accepted as an order, through all the various other departments, sales, development, production and distribution. It does not follow, however, that it is the prerogative of the estimating department to do the controlling. This is outside the function of the estimating department. The manner in which this effects the estimating department is whether there is any feedback of information or not from these later stages. Feedback at all times is essential to the development of the estimating department; whether it follows immediately after each operation or at a later stage, when there *must be* a comparison made between the estimate and the actual final cost. This is where management can check upon the validity of the estimating department and its worth, and from this formulate policies and directives for the future of the department. Although at this stage the accuracy of an estimate can be assessed, it must be compared like with like, so that the limitations and inaccuracies are valid. A common occurrence is that the estimate was compiled six months before its acceptance and possibly a further six months has elapsed before any production, after which there is a strong probability of there being a difference between the two. When this occurs there are many possible variables that could have taken place; modifications to the original, the cutting of overhead, labour or material costs by another department (who are prepared to take a calculated risk;

for example, on the assumption that the estimated production figures are too low, and it will be possible to reach higher production speeds). Estimating is therefore a positive stage within the functions of a company. It does not and should not work alone; it is as dependent upon the functions of the other departments within a company as they are upon the functions of the estimating department.

A good management expects to receive a functional result from the estimating department which is in accordance with the overlying capability of the company. There is adequate evidence that estimating need not be slip-shod, but that it can become an exact art, not necessarily a science, a fact that should be accepted upon a governmental scale as well as upon a company scale. (The results obtained from government estimating is all too painful to the taxpayer, with all the continual escalation of taxation.)

Management, therefore, can only expect to receive from the estimating department, finalised estimates of an accuracy which is totally dependent upon the experience and training of the estimating staff. Many managements in the past have been able to get away with a low-grade estimator for the low state of accuracy that has been required. With modern financial requirements, this has gradually been swept aside as more and more companies find that the competition is getting hotter and there is a need for a much higher state of awareness and accuracy.

As management advances its own techniques, it reverberates into the estimating department. It is therefore necessary for estimating techniques to keep up-to-date with these. This can only be achieved if management understands the meaning and scope of the estimating function.

# CHAPTER 2

# DEVELOPING FORMS AND TYPES OF ESTIMATES

Each industry will develop the form of estimating that is most conducive to its own environment. To this end it can safely be assumed that upon receipt of an enquiry, the form of the estimate will be based upon assumptions that can only be arrived at according to the knowledge available at the time, and the time available to compile the estimate. It is not generally appreciated that the time factor can be as critical as it is, particularly time/cost ratios and the time that an estimate is received. Primary and cardinal errors creep in at an early stage, and by the time they are rectified, or attempts made to rectify them, it is too late to be accurate. Later, when differences show up between product costs and the estimate, the fact that they may have been caused by a functional breakdown or political expediency is often overlooked. There are, however, seven major causes for differences between product costs and the estimate. These are:

1. Changes in detail by the customer, with the subsequent acceptance of the order, without checking with the estimator regarding the full significance of the change.
2. Lack of appreciation of the specifications.
3. Production changes in the method of manufacture, or the planned for machines being unavailable.
4. Changes in labour rates, material costs, overheads or profit rate.
5. Unexpected technical breakdowns.
6. Treasury insistence that estimates are based upon current rates and prices.
7. Lack of adequate staff in the responsible technical branches in the months immediately preceding the placing of the project study contract and also during the project study itself; alternatively, projects discussed at directorship level without consultation with the technical branches.

Over the years, estimating has developed into six major forms which are understood by most estimators, although no clear-cut definition has been laid down. These cover the subjective, parametric, comparative, synthetic, global and research and development forms of estimating; the late Ministry of Technology in actual fact came out in favour of these titles. A fuller description of the forms of estimating follows.

## Subjective Estimating

In the subjective method of estimating, the estimator obtains his price (for a project or unit of manufacture) by relying completely upon his own memory and/or judgement, sometimes supporting this by referring to his own personal notes collated over the years and which have probably been kept in an illogical and haphazard manner. At no time while compiling his estimate does the estimator refer to systematically collated data which have been collated and analysed from previous work. This, of course, is a totally unsound and unsatisfactory method of estimating, although at times it is often used as a quick method to check a price or give a customer a rough guide to the general area of a price. This method of estimating is unfortunately widely used throughout industry. It cannot, however, be a method which can be at all recommended. It is totally unsafe and should only be used as a last resort or for the reasons previously stated, i.e. a quick price or a guide price.

## Parametric Estimating

The parametric method of estimating is widely used in the petrochemical, chemical and related fields. It is a method of estimating the labour and material content of a single technical or physical characteristic or combination of these of a past project which has been statistically analysed.

Having statistically analysed the related costs, these are then expressed as a generality to a characteristic, from which a formula is derived. For example, the output volume of a plant is established as a cost relationship to the cost of a plant of a particular size. Although not an extremely accurate form of estimating, it does, however, give a cost accuracy for a plant which is near enough for all practical purposes, upon which the board of directors can make a decision.

## Comparative Estimating

By means of comparison of like with like, based upon the comparison of part or the whole of a project or component with formal records which have been kept on past work, a price is derived. In addition to this comparison, allowances should be made for any differences in type, scale, complexity or any advance in design or technology. This form of estimating draws its assessments from past experience rather than to make a detailed breakdown and analysis of component functions, with the additional analysis of work and operations to be done, labour, materials, overheads and any other costs or charges.

[*Note:* A comparative estimate in practice could be made at the same time as a parametric estimate, the essential difference between them being that the comparative estimate is arrived at by expert knowledge of the cost relationships which are likely to take place between the chosen comparative projects or components, whereas the parametric estimate involves the applications of a formula to selected parameters, applying a formula which is judged to be appropriate to the particular case.]

## Synthetic Estimating

Synthetic estimating can be either broad or detailed, usually dependent upon the time available to the estimator or the value and/or complexity of the product or project. Its aim is to break down the job into its smallest constituent tasks based upon the best available information at the time. The value assessment at the time is carried out on individual tasks, the speed of operations and/or work, labour costs, material costs, availability of plant, variances, etc., required for the task. The total estimate is made up of all individual costs of the various tasks or operations and any additional allowances for any assumed or anticipated extras or increases in the costs of material, labour, etc., or technical difficulties. A correctly laid out synthetic estimate can and is used in conventional engineering techniques as a production layout sheet, as it is usually possible to obtain a fully detailed design drawing or set of drawings upon which to base a synthetic estimate. With the full knowledge of the processes and materials involved, it is possible for the estimator to make a full and complete analysis of the various stages of production and the estimate can be built up into a complete and factual document. The depth into which synthetic forms of estimating can reach is entirely dependent upon the degree of detailed design available and in its broadest sense can, for components and sub-components, be arrived at in areas of work by comparative and parametric methods. Conversely, with full details and information the synthetic method can and does arrive at an extremely accurate estimate. Although estimates of extreme accuracy can be achieved by this method of estimating, the depth into which to go may have to be curtailed as the economics of the estimate overshadows the cost of the job.

## Global Estimating

Often a quick subjective estimate is required in order to ascertain whether it is advisable to continue with a project or not. In this context the global estimate can be justified; they are primarily estimates which are made in overall terms based upon past experience, so that a quick subjective analysis of a problem can be made. They should therefore be treated as such and not as a fixed price estimate.

## Research and Development Estimates

It is well known that the field of research and development presents one of the greatest difficulties in respect of estimates. Basically there are two methods which cover research and development. The first is a simple form of presenting research and development with a target to be reached within a fixed budget. Secondly to estimate for research and development costs. This form of estimate is usually arrived at by a series of stages through trial and error that can only be overcome by full and frank interchanging of ideas and information, including references to all available records of past research and development work and personal past experience in the preliminary stages. Because of this the essential point of making allowances or tolerances (not contingencies) to cover any foreseen or unforeseen difficulties which could or could not arise are often left out, these being one of the major reasons of under-estimation. The variances which occur in this field make it necessary to have skilled and trained estimators able to analyse this form of work.

Having defined the area to be estimated, it should be possible to use one of the major forms of estimating, although in the various sections of industry, estimates are known by different names. In the Glossary at the end of the book you can find some of these definitions.

Management should therefore be clear in its own mind as to the form estimates should take within its company and if necessary these should be clearly defined so that all within the company are aware of the accuracy referred to when relating to a particular estimating type. For example, it may be the policy of the company to do estimates for a preliminary stage to an accuracy of between 20% and 25%, a second stage between 10% and 15% and a final stage upon full details of between 2½% and 5%. If this should be the case, some form of nomenclature should be available within the company to easily define the type of estimate used. It cannot be stressed too greatly that the formulation of the individual forms or types of estimates is the prerogative of the company and the particular industry. There can be no hard and fast rules laid down, as these will alter from company to company according to the importance given to the estimating function. The development of the estimating function will be dependent upon whether, for instance, it is possible to estimate synthetically to a degree of less than 2% accuracy because of the completeness of the information given in an enquiry.

It has often been said that one cannot get an accuracy of below 2%, but this can be achieved if estimating is correctly formalised. This also is applicable to major capital projects. These can be controlled and brought to within 1% of the estimated or budgeted cost, but this is dependent upon the decision of management to formalise correct estimating techniques, followed by an efficient cost control. This is possible, within any company, provided management accept the new breed of engineer, a fully trained cost engineer who is capable of dealing with all aspects of cost engineering.

A more detailed description of developing estimates is covered in later chapters. We have at this stage purely and simply looked at various forms of estimating. I have also stated that accuracy can be achieved, but that this is dependent upon the overall teamwork that can be achieved within a company by having laid down procedures and policies to follow.

These policies and procedures are in themselves dependent upon the ability of management to formulate what direction the company is going to take, its corporate strategy and its future technological forecast.

# CHAPTER 3

# BASIC STEPS IN ESTIMATING

Before one can attempt to digest the basic steps of estimating we must first define what is an estimate. The general definitions which have become accepted for an estimate are as follows:

(a) Any approximation of conditions, past, present, or future. The level of the variable which is believed to have a 50% chance of achievement. In this context, estimates of fixed investment are defined in the following order of increasing accuracy: order-of-magnitude; study; project control; firm.

(b) A quotation figure forecast to limits of accuracy compatible with the use to which it is to be put and the policy decisions at stake.

(c) A forecast made in advance of the expected cost of a project.

(d) A compilation of quantities and costs of all components of a defined project.

(e) To judge and form an opinion of the value of; to compute; to calculate; to offer to complete certain work at a stated cost.

Of these definitions or concepts, I would prefer to accept the last definition — to judge, to form an opinion of the value of. This starts to give an indication of what estimating is really about. The general supposition that an estimator simply takes out his little black book, extracts what information he requires, adds a few other meaningless figures, adds in the date, multiplies it, puts in his contingency, looks at the figure, locks his black book away, scraps the figure he arrived at, guesses at a new one, with which all disagree, then falls asleep exhausted, is a thing of the past.

An estimator today must be able to judge. Through his judgement is the ability to assess. After assessing to the degree of accuracy required, he must be able to form a final opinion as to the actual nature of the work to be done, how this work is to be done, where and why, followed up by computing and calculating his assessments and opinions in order to arrive at a final cost. It is no mere guesswork. If it was, then I am afraid that 90%

11

of the engineering industry would be bankrupt, a fact which could be proven, but which has not yet come home to roost in many a boardroom, and which remains hidden in middle management. The basis of all engineering estimating is, in the beginning, an adjudication of alternatives by the estimator. He must therefore have guide lines to follow. These guide lines are the procedures and policies laid down by the board of directors for the estimating function, where is the company going and what is it going to do? If the future of any company starts anywhere, it is here in the estimating department. Enquiry, quote, manufacture, delivery, then repeat if the customer is satisfied. No order, no future. The format that an estimate has to follow has partially been laid down by the nature of the industry, then by company policy and directives. From here on the estimator has a base upon which to work. Unfortunately some management decisions have a disturbing effect upon the work of the estimator. The reason behind this is normally the extent to which the estimator is kept ignorant of any managerial decision until after the estimate has been compiled. When this occurs it is primarily because of the out-of-date attitudes of management, particularly where management look upon the estimating function as one of a necessary evil. On the other hand, it is merely ignorance of the estimating function or an oversight of the true function of estimating. It is only when the estimate has been prepared ready for quoting that management becomes aware of the fact that some policy decision has not been adhered to. When this does occur it is rapidly brought to the attention of the estimating department. The following effects are two-fold: one of annoyance in the department at being considered too mundane to be informed of such matters, and secondly the tremendous amount of time that this type of occurrence wastes, besides focusing attention on a particular job which then has to be rushed, sometimes with disastrous results. It is important that the estimating department is kept fully informed and abreast of all necessary company policies and decisions, and a good reason why the estimating department should be placed under the control of the Financial Director (see model, Fig. 1), as under his direction these policies can be implemented to the best advantage, without any control interference from other departments. Too many times it occurs that when the production or sales departments have the estimating department under their control they alter estimates to suit their own purpose, not keeping them factual for the full benefit of the company.

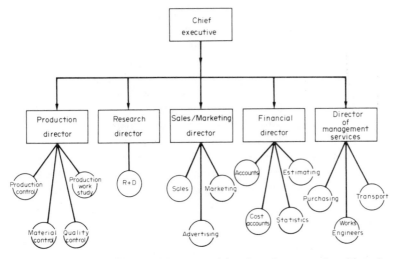

Fig. 1. *Typical management organisation model* to show the suggested position of an estimating department in the organisational pattern.

The first evaluation that the estimator will undoubtedly make of the job on hand will be to divide it into two categories:

(a) The work to be done by our own company.

(b) Work to be done by outside companies.

After this he will judge the actual amount of estimating that is to be done and approximately how long it will take him, bearing in mind the length of time that it will also take to obtain any necessary outside quotes. When he has decided how long it will take to prepare the final estimate, he will then compare this time against the time by which the final estimate is needed. Now the customer or sales should be informed if the estimate is going to take longer than the required time, not on the day that it is required, as informing the customer, particularly government departments, that you require an extension of time normally means that you obtain the extension. This means that your quotation will be considered. Leaving it until the last minute has the opposite effect, and in most cases your quotation is not accepted. In many companies management lay down a policy relating to time, i.e. that all enquiries of a general nature will be prepared within a prescribed number of days. This gives the estimating department a defined limitation, which also gives them a base for the depth of estimating detail. The detail they require for each level of limitation is of this or that amount and is of an accuracy that is acceptable by management.

Preparation time for estimating is of paramount importance, as upon the length of time given to prepare any estimate depends the accuracy of the estimate. Policy must lay down the depth of detail required for any type of estimate. In some work the actual timing for the drilling or piercing of a hole will be required; in others the approximation of the cost of the whole of the bracket will be near enough, taking into account the overall cost of the job and the economics of the estimator having to work out all the details. One must become aware of the fact that there is an economic level when it does not become viable for the estimator to be employed working to a greater level of accuracy. This level is well known within individual estimating departments.

### Special Areas

At this preliminary stage the estimator will also pick out any area which is of a "special" nature. These fall principally within four categories:

1. Special materials, i.e. specific specifications, extended deliveries.
2. Development tests.
3. Redrawing.
4. Design work.

All these present difficulties, which can hold up the preparation of an estimate; for example, particular material may also require development tests; drawings may need to be redrawn so that the article can be manufactured. (This occurs to a great extent in the foundry industry where drawings are sent to the foundry by customers to engineering standards, not foundry standards.) The experience and training of the estimator will enable him to separate these items from others and he will be able to take the positive action required, such as getting outside enquiries off as quickly as possible, or making the drawing office commence work. The fundamental essence of any estimate lies in the natural breakdown of a job into its own particular logical sequences. These will fall into a logical estimate, regardless of the size of the job to be done; an estimate for a car component is treated basically the same way as a project for capital plant. In all estimates the essentials are worked out and a rough estimate made. This is tidied up and the estimate finally becomes the basis of a quotation to a customer, or to the board for authorisation of a project. Owing to the complex nature of estimating, every estimator in his own field becomes a specialist, and although estimators have this tendency to become highly specialised in their own particular industry or field, the background work is basically identical. Every aspect of the project has to be analysed to find out what part it plays in the overall picture. A hole or a slot in a bracket can have a similar significance in the overall pattern of an estimate as does a surge hopper in a process layout, the difference being in size and complexity. For example, to an estimator the only difference between 1,000,000 and 100 is four noughts. He only has to assimilate the larger number in his mind and work accordingly. The same general logic applies whether one is estimating for batch quantities of articles costing 5p each, £1.00 each, £100.00 each or more, or even a capital project for half a million pounds or a government defence project for 100 million pounds; one simply applies the same general logic in the larger fields as in the smaller. (In actual fact in the larger fields it is an estimating team rather than an individual, who are specialists as against an all-rounder.) Here there is a basic approach that can be made to any estimating problem, which becomes easier as experience is gained and the nature of the special areas evolve within one's mind; when a state of awareness as to their interactions becomes a natural instinct.

A basic approach presupposes that estimating has a logical plan. It will not come as a surprise to the reader that estimating has a seven-point plan which can be followed in any basic approach to a given job. If an estimator follows these basic seven points in his initial evaluation of any job he will very seldom ever make a fundamental mistake or error.

### The Seven-Point Plan

1. Have I all the relevant papers, documents, drawings, specifications, etc., that I require in order to understand the job?
2. Into what logical steps do I need to break the job down, into what detail shall I go, and what is the time available?
3. Are there any significant specifications or tolerances?

4. Do I require any quotations, data or information from any outside source?
5. Does any point require clarification, redesign or checking for manufacturing difficulties (i.e. can we manufacture the component as drawn)?
6. Have we the capacity to manufacture the components required, or can the plant be put where shown?
7. Do I require any further advice or data (synthetics) for the elements of the estimate?

Having covered these points, the estimator can go ahead and collate all his facts and figures, prepare his estimate, stating any limiting conditions, his final price and the type of estimate.

As these seven points are important, let us examine them more closely:

### Point 1

Have I all the relevant papers, documents, drawings, specifications, etc., that I require to understand the job?

### WELL, HAVE I?

One practice within a manufacturing company is for the estimator to be asked by the sales department for a price for such and such a component, quite often only being given the last order number. It is also a common fact that the sales department is only interested in obtaining a price as quickly as possible. In wanting this, they have an extremely bad habit of passing only what information that they think is relevant and necessary. Also, being in the majority of cases completely non-technical they are unable to recognise whether any important details are missing or not, or if they do know, in their eagerness to obtain an order they overlook details, which could, in effect, turn out to be major disasters. Many companies have, in order to remedy this type of occurrence, introduced preprinted enquiry forms upon which the sales representative, clerk or customer can enter all the relevant details, measurements, sizes, etc. Unfortunately these have limiting factors, which in general only allows them to be effectively used where components, articles, etc., have common points (points which either confirm the component's size, power, etc., i.e. an electric motor; a steel joist; or alternatively the chemical make-up or formula of material).

There is also the tendency of the sales staff holding onto documents and papers unnecessarily, either thinking that they are superfluous to the enquiry, or that they may get lost in transit between departments. When the time comes that the estimating department is aware of the existence of these additional items, it is usually too late to make any alteration that may be necessary. The estimate has been finalised and quoted; one then has the formality of going to a customer to admit to a faulty quotation, or accepting the fact that there is a high probability of a loss being incurred by the company. If the former action is followed, it gives a bad impression to the customer; having either quoted too high or too low, they wonder what type of company are they dealing with, and lack confidence. Occurrences like this should be totally avoided, and a well-laid-down estimating procedure would make sure that this could not happen. Ideally, of course, the sales department should record all the documents received, acknowledging their receipt to the customer, either passing all the documents over to the estimating department or having photostats made. It is, of course, vital and of considerable importance that the estimator obtains all the relevant details and documents regarding an enquiry, particularly where there has been any preliminary talks or discussions between representatives of the company and the customer. Should this be the case, these preliminary reports often contain vital information and can also give a historical background to the work, which enables the estimator to become more conversant with what is actually required. Additionally these reports may contain some deviations from previous company policy which may affect the job, and the estimating department can let management know if these policy variations will have any side effects upon any other work. This may seem to be straightforward and plain common sense, but in actual practice procedures seem to go astray as certain individuals force their own will or personality upon other employees. This is why it is important that there should be a well-defined estimating procedure and policy laid down by management. It may seem trivial, but it is upon the accuracy and reliability of any forecast given or emanating from the estimating department that any ultimate profit of the company is made. An informed and reliable estimator, if he has had sufficient experience and training, should be able to arrive at a solution or price which:

1. Enables an order to be obtained.

2. Enables work to be done by the company and at a profit, or alternatively allows a contribution to the company's income to be made.
3. It can be done at a price and quality which is acceptable to the customer and also is at a price which enables reasonable wages to be paid.

Any figure or price which is the direct result of the estimator's deliberations considered to be necessary in order to obtain the qualities of 1, 2 and 3 should be based upon as factual data as it is possible for the estimator to obtain. In order that he can estimate accurately, any attempt by any other department to sway his judgement so as to arrive at a predetermined figure should be resisted. What happens to the estimate after it leaves the estimating department is the prerogative of the sales department and management, but in the initial stages the estimating department should be made aware of all the pertinent facts relating to any particular job; not to have these is to make the whole thing a charade.

*Point 2*

Into what logical steps do I need to break the job down, into what detail shall I go, and what is the time available? All these naturally are interrelated one to the other; if we have the answer to the third part, we have the answer to the second, if we have the answer to the second part, we have the answer to the first part.

Where do we start? As it is more usual for the time factor to be the most important of these, we will deal first with the question of the importance of time being available to the estimator. Time available? What a point! How often do we get asked for a price yesterday, not today, not tomorrow, but yesterday, to which the old saying seems to be the answer, do you want it Thursday or do you want it good? How prevaricating it gets when one is asked for a price immediately, only to find from the documents that someone has been sitting upon them for anything up to a month or more. It is this question of time being available, time in which to do an estimate, which is of vital significance to the estimator; the right length of time being available makes all the difference between an accurate or a slip-shod estimate being made. Time being available, however, does not mean all the time in the world, but sufficient time in which to do the type of estimate asked for, time in which to prepare a courageous, forthright, factual, realistic estimate. When time is short, the estimator becomes vulnerable; too many things have to be done too quickly without much thought, resulting in figures which can only be described as airy-fairy pipe dreams.

Each and every little point or aspect of an estimate can have a considerable amount of time spent upon it according to its importance in the overall pattern. A considerable amount of time also has to be spent upon the assimilation of knowledge relating to all the salient points of an enquiry. This aspect of the estimating function is not clearly understood by outsiders. How often does one hear the remark "what asleep again" being made to the estimator, when all the estimator is doing (probably with his head in his hands) is looking at the blueprint or drawing and quietly attempting to interpret this into a three-dimensional image in his mind.

The interpretation of a job on hand into a three-dimensioned image is not always understood; most estimators have the ability to interpret any drawing into a three-dimensioned image in his mind, and in doing so understands the problems more clearly and has a better idea of what has to be done. All this requires time, reasonable time, so that the estimator can assimilate all that he requires from the drawings and documents on hand.

Unfortunately this aspect of estimating is not always appreciated by management, although it is one reason why the estimating department should be placed within a quiet area of the offices or works, so as to be able to work without any distractions. The fact that the time aspect of an estimator's function is not clearly understood can be illustrated by the following case. A well-known firm of management consultants called in to evaluate and streamline the office procedure also extended this into the estimating department; after many months it was announced to the estimating staff that in future all estimates had the following time limits: a building estimate was to be done in 15 minutes, a general engineering estimate in 20 minutes and a motor vehicle component in 25 minutes. All these were totally unrealistic and it was assumed that the consultants were having a joke at the expense of the estimating department. The staff, however, were assured that this was not so and that these figures were to be adhered to. Fortunately an enquiry came in at this time which took over half an hour to merely check the drawing numbers against the enquiry. This showed the total absurdity of these times, which management then dropped.

However, there is no reason why time limits should not be set for estimates, but these limits should be related to the enquiry on hand, its complexity and the type of estimate required. As the form of the estimate and the type of price required will determine the time factor, this should be agreed to between the relevant departments; a quick budget price, for instance, can be agreed to far quicker than a fixed price estimate. The determination of a type of estimate does not only give an indication as to the time available for the estimate, but the time available gives the estimator an indication of the depth he can go into for the individual elements. It is important that anyone connected with an estimate realises the significant part that an appraisal of the drawing takes in relation to the time allowed for the estimate. Many outsiders see only the final tip of the iceberg, that is the final sheet of the estimate giving a final price.

It is all the underlying workings and factors which have to be appreciated, the thought, the various ideas and alternatives, the calculating and computing the final solution, this is what is not seen. The amount and complexity of detail to be worked out, which will in turn be controlled by what detail has to be achieved for the type of estimate required, whether it is a quick budget price or a quick appraisal of all the major items and their costs or a detailed synthetic analysis for a fixed price estimate. The first two do not take too much calculation, merely confidence in one's own ability, based upon previous experience, whereas the fixed price is much more complex.

What is the market going to be over the next twelve months or possibly two or three years; in some cases it may be up to five years, all of which gives the estimator the added responsibility of having to make judgements extending over a lengthy period of time, when having to split up the various component parts of an estimate will have its effect: a motor car component having every piece of material, every single operation, pressing, cutting, rolling, drilling, welding operation, etc., all worked out to its smallest element. Capital plant, on the other hand, would have to be split into functional section costs, most of which would be the subject of quotations received from different manufacturers; the function of estimating, therefore, has another facet: are you a manufacturer, assembler or a process buyer? This facet again is a limiting factor upon the estimating department's function which will determine the type of estimate to be made. The nature of the industry within which the estimator is working will help him to determine the logical steps to be taken in formulating the estimate. The complexity of the job will also give an indication as to the separate component parts of the estimate. Whether there has to be sub-estimates or not, this will all be determined in the initial breakdown, and whilst making his initial appraisal the estimator will jot down any ideas or questions to be asked. In retrospect he will be able to know whether, in fact, he will be able to do the estimate in the allocated time or not.

*Point 3*

Are there any significant specifications or tolerances? This is another important phase in the development of an estimate, as upon this will depend the ability of the estimator to be able to recognise the importance of any specification or tolerance given. A considerable number of specifications and tolerances are normal working practices. Being able to accept these as such, and being able to appreciate that others can present difficulties, is a salient point. The estimator must have the ability to recognise these so that he can save the company headaches, or even considerable sums of money. Some specifications and tolerances, by being accepted without thought, can mean special materials and/or special manufacturing techniques, all of which involve added costs. It is not good enough to be able to recognise that a component has been made previously, therefore it can be made again; it could be that the specification, tolerance or finish has been changed; in these cases the price cannot be merely uplifted, it must be re-examined and repriced if necessary. This applies particularly to components being manufactured for the automobile industry, where the suffix letter of a drawing could have been altered from 'A' to 'B', or 'B' to 'C' WHEN THIS COULD EITHER MEAN left hand or right hand, or an alteration in the type of material (mild steel to chrome iron or aluminium) or the finish (from untreated material, to painted, chromed or anodised). Nothing can be taken for granted; each specification and tolerance must either be known or be checked; it is better to be safe than sorry. There should be no difficulty in the estimator being able to find out the meaning of any specification. There is a complete wealth of knowledge within companies and associations, who are only too willing to be able to clarify and give all the necessary help to anyone requiring knowledge about any specific specification relating to their industry. It is also unnecessary for a small company to employ a metallurgist full time when his professional advice is only required a few times a year. An estimator should for his own benefit make sure that his own or the company's technical library carries all the standard specification handbooks relating to his industry, in order that he has immediate access to the most common specifications, references and tolerances. A trained estimator will by his own expertise automatically know what parts of an enquiry should be checked and double checked regarding specifications and which areas will require special attention during the estimating function.

*Point 4*

Do I require any quotations, prices or information from any outside source? By the time we have reached this point our notes should already have given us the answer to this question, as the vital point in this respect is the time it takes to obtain an outside quotation. First the estimator has to make a draft of the enquiry to be sent. This has to be passed on to the purchasing department, who in turn sends it out to prospective suppliers, generally to three in order to obtain comparative and/or competitive prices. Owing to backlogs and priorities in work going through the purchasing department, these enquiries are at times subject to heavy delays, which only means holding up the estimate or having a "guess" at the price in order to get the quotation out. Sometimes special material requires complicated and protracted discussions; one such problem in the past was that special steel was required in order to manufacture the stiletto heels for women's shoes; tomorrow it may be that carbon fibre could be the answer to a particular problem. In many of these cases the company sending an enquiry in has no realisation of the length of time or the cost involved in carrying out tests on materials, or the difficulty in finding a suitable supplier of a special material. So many enquiries of this nature have only been sent out by clerical staff who think that it would be a good idea to obtain related prices, without having the faintest idea of the many problems involved; when this type of enquiry is received it is a good practice to contact the company concerned to ascertain whether they are aware of the associated problems. In some cases the company is well aware of the facts and is only too willing to not only await the outcome of any tests, etc., but also to pay for any costs involved. Specific tests and developments which could be carried out in various industries cost a great deal of capital expenditure; this fact must always be in the back of the estimator's mind, that he must not accept work of this nature too readily. Sometimes it is easy to have some form of development work carried out with little cost. This occurs in material finishing, where one can obtain another component which is being manufactured in the same or similar material and have this processed in order to obtain a sample of the finish to see whether this would satisfy the customer. On another occasion it occurs that in order to manufacture a component the work has to be done on a machine with a very high work load. This results in having to give a protracted delivery time which is known to be unacceptable to the customer. (In these cases it may well be desirable to evaluate the possibility of farming the work out in order to obtain the order.) The question of a machine having a high work load or lead time is one which the estimator has to watch very carefully, as it is useless to obtain work if one is unable to meet deliveries, and subsequently have orders cancelled. This, however, is a digression into point 6, other than the fact that if we farm work out we should get any enquiry off as quickly as possible.

## Point 5

Does any point require clarification, redesign or checking for manufacturing difficulties (e.g. can we manufacture the component as drawn)? Many problems relating to manufacture do, however, start in the drawing office stage. The way in which drawings have been prepared is of major importance. Drawings can reflect the capabilities of a company's drawing office staff, when we could ask ourselves whether the drawings are being prepared by qualified draughtsmen, apprentices, or merely being drawn or sketched by an office clerk? It may be that they are being prepared by a tracer, a designer, or even the managing director himself. No matter who has prepared a drawing, the question is, has it been done in a capable and professional manner? A drawing is only as good as the training and the experience of the person preparing it. Some companies have a high reputation for their drawings, to the extent that if anyone thinks there is an error on the drawing, it is checked and rechecked to make sure that it is not our capability of interpretation that is wrong. Bad drawings, or alternatively the inability of the estimator to be able to read more than the simplest of drawings, can lead to errors. Any doubt, therefore, about a drawing should be dispelled as quickly as possible, and in many cases the words "If in doubt ask" are clearly printed on the drawing, and it is a wise man who follows this advice. Clarification of a drawing should have a priority commitment, and until all the points on a drawing are clarified it is rarely possible for an estimator to begin his work. Along with the clarification of a drawing is the accompanying fact whether a simple redesign would have any significant effect upon being able to reduce costs. An estimator can, at times, see that an alternative manufacturing process or an acceptance of a nominal size hole would mean standard tools, thereby saving tooling costs. Draughtsmen of one company cannot be expected to know the manufacturing processes and tools of another company, therefore should there be any possibility of a redesign, the draughtsman responsible for the drawing should be asked if possible whether this could be allowed. As this has, in most cases, the effect of keeping costs down, most companies are normally only too willing to have a look at any possibility of a redesign.

It is the ability of the estimator to be able to translate drawings clearly which is an important factor, and why training should be orientated towards drawing technology. In addition, the personal experience of the estimator in shop floor and work shop practices has a considerable bearing upon the validity of the estimate. Without a full understanding of how and why labour and machines work, the estimator cannot even commence to estimate.

*Point 6*

The understanding of point 5 takes on considerable importance when one has to decide whether we have the capacity or the ability to manufacture a component. For a company to be saddled with manufacturing a component which is totally alien to its production potential puts the company in rather an awkward position. Whether we have the capability or the ability to manufacture will already have in many instances been agreed to within the parameters of the previous points, and by this time it will be principally borderline cases which will need further discussion between the various other departments within a company. It may well be the question of tooling, or it may be a production difficulty; all will have to be ironed out by discussion. Some enquiries may mean a diversification from the normal run of the mill, but still within the scope of the company. It would not be practical merely to turn an enquiry down because traditionally we do not make the component asked for. With a little thought and probably a little outlay of capital expenditure it may be possible to manufacture and be another source of income. Estimators should be flexible enough to be able to recognise any of these possibilities. Because we are traditionally a kitchenware manufacturer does not mean that we cannot manufacture a motor car component, should a suitable opportunity arise. All these questions would be answered by the particular resources that the company has. No two companies are the same, the size of the buildings, services available, layout, machines, labour, etc., all have their own individual interactions and so will raise their own particular set of problems, all of which the local estimator will know.

The estimator should be kept informed of all the work loads of the different machines, the conditions of the machines. Have they broken down? If so, are they under repair? If under repair, how long is it going to take? If not undergoing repairs, why not? When are repairs going to commence? Or are the machines to be replaced? If the machines have been moved, what part of the factory have they been moved to? It sometimes happens that moving one machine to another department ensures that the machine is kept in constant use, and at the same time makes it unavailable for new work. To keep the estimator aware of all these occurrences does not automatically mean additional staff in order to see that this type of information is made available, as a short perusal of all the existing means of communication and paperwork will find that all this type of information is already being passed from one department to another under normal working functions. What it means in most cases is simply another piece of carbon paper and the estimating department being put upon the circulation list, without any additional work. For example, the buying office knows when new machines and raw materials are being bought, and if a stock control department is in existence they will be doing returns upon the stock positions. It would also be a simple matter for engineering or maintenance departments to inform the estimating department of the likelihood of any protracted repairs having to be undertaken, as well as the progress or production departments passing on copies of their machine loading tables. This all returns to the necessity of a company working as a team, not as separate segments working for themselves.

*Point 7*

Do I require any further advice or data (synthetics) for the elements of the estimate? This is where we come to the end of our preliminary investigations into any enquiry; here is where we collect all our thoughts and ideas together and our estimate begins to take shape. As we lay out our estimate we evaluate all our separate elements and in doing so find out whether we, in fact, can complete the estimate or do we need further discussions or facts to be able to finalise our figures and work out our final calculations.

All these seven points follow a logical pattern, which should follow a preconceived plan for the type of estimate to be compiled; in fact an estimating plan should be laid down in an estimating manual for the company (see Aids to Estimating). The principles underlying the estimating function are simple. They are a series of logical points to be covered so that a problem can be broken down into smaller tasks in order to make them manageable and able to be solved. Having broken down the major task, one proceeds through a series of sequences, probing and resolving any difficulty, until at last an estimate is formulated and it is possible to work out the final calculations. Our next chapter deals with some of the individual estimating elements, and an example of how to work out an estimate for an automotive component follows in Chapter 6.

CHAPTER 4

# ENGINEERING ESTIMATING ELEMENTS

Having worked through the basic steps of estimating, we come to a point where one is ready to lay out a preliminary estimate, detailing all its individual elements. In engineering there are many of these basic functional elements which must not be confused with SI units. This chapter deals with the most common of these. SI units are chiefly concerned with units of measurement or modules and in engineering it is an everyday occurrence to manufacture parts which later form part or whole of an SI unit but because of design, etc., are in fact bastard in practice and do not conform to a modulus unit. In engineering estimating one is therefore concerned with elements of manufacture relating to measurement in the metric system only.

## Materials

The choice of material has normally already been decided upon by the customer. He has in his enquiry stipulated, by a specification, what type of material he requires and whether he requires alternative prices for various materials or finishes. The supplier has satisfied himself that the material is suitable for the job to be estimated and that the job can be made from the specified material, although this is not always the case.

In engineering, materials can be classified into two groups:

(a) Metallic materials.
(b) Non-metallic materials.

The metallic materials are sub-divided into ferrous and non-ferrous, and common non-metallic materials are plastic, timber and rubber. Most materials are held by companies (in stock) according to individual requirements of each company, principally governed by the particular industry in which they are manufacturers. Some industries will only stock coiled strip, others flat steel sheet, others wire and rod, whilst others will hold scrap steel and pig iron, or any mixture of one or more forms. The variations can also be in different types of material, brass, aluminium, steel (mild or stainless). From these few indications the variables will be seen to be endless, before one even tries to think of the thickness or gauge of material.

The manner in which a company holds its stocks of material suitable for the industry it is in will be a determining factor upon whether it is able to compete with the price in the market or not.

The principal material used in engineering is H.R.P.O., H.R. or B.C.R.S. steel sheet or strip.

H.R.P.O. (hot rolled pickled and oiled). Is suitable for welding and painting, but not for electro galvanising or plating purposes.

H.R. (hot rolled). Being suitable for sections of a simple shape, it is however unsuitable for resistance welding unless of course it is first descaled; it is also more difficult to protect against corrosion owing to scale.

B.C.R.S. (bright cold rolled steel). Has the accuracy of profile which can be obtained to the order of plus or minus 0.25mm; it is also suitable for all types of welding and nearly all types of protective treatments including plating.

It will be appreciated that the estimator must not only keep in touch with the stock control clerk, but also with the buyer, in order to keep up-to-date with the stock position, its availability and price.

## Scrap

Scrap in engineering can be accounted for in two basic forms:

(a) Scrap material which can be accurately worked out as that portion of material which is going to be a definite loss through the method of formation or manufacture.

(b) Scrapped material through faulty or bad manufacture or work-manship or subsequently faulty material showing up in the manufacturing process.

An example of (a) would be scrap material surrounding a pressed blank, whereas (b) could be laminated steel.

## A Materials Element

A materials element of an estimate must be that proportion of material which is not only necessary to manufacture the component, but also the amount of material that will be actual scrap. (All actual scrap can be legitimately charged for against the cost of the component.) There are, however, times when actual scrap can be used for making another component when allowances can be made for the material, if feasible. All scrap material has a value, but where WASTE is concerned this has no value to the company. It is a complete loss, although there has been an initial cost to the company. An example of waste could be machine oil used in production, where there is an actual percentage of oil being continually lost without any recovery whatsoever.

## The Press Element

The press element that an estimator uses is 'X' number of operations per hour. This takes into account not only the speed of the press and its power but also all handling and clearance times. An estimator has to make a decision, "What press is the work going to be done upon?" The major part of work and manipulation carried out in engineering is performed upon power presses, hand presses and brake presses, and all forms of operations which are carried out upon presses have a basic element which is dependent upon the form or shape that the work takes. Some are merely straightforward piercing operations, whilst others are a series of operations to form or manipulate a particular shape. The type of press upon which work is performed is dependent upon the actual power required to carry out the operation; one has also to decide whether a single, double or triple action press is required. The power required for an operation can be worked out by a mathematical formula, and once the type and size of the press has been determined, one estimating factor has been set, i.e. the cost of the press overhead. overheads of presses are worked out by the accounts department (as are all other overheads), each press carrying its own overhead. These overheads are given to the estimating department by the accounts department at predetermined intervals, and when costs escalate at intervals in between they are updated. It is also usual practice for all overheads to be set by the accountants without any consultation with the estimating department.

Press overheads are calculated by the hour, to which is added the relevant labour rate; the method of setting overheads varies from company to company dependent upon the manner in which accountants are reclaiming capital costs. They are, however, always set out in a logical manner which is uniform for the recovery of overheads for any form of machine. In many cases, presses are first sub-divided into their particular category, then into size, normally stepped up in capacities of 5, 10, 20, 25, 50, 75, 100 and 150 tonnes. Presses of a higher capacity are normally to be found only in heavy engineering. An important fact to remember is that to put a component upon a 75-tonne press when a 50-tonne press would have been sufficient does not only inflate the overheads and possibly lead to the loss of an order, but it is also a loss of capacity within a company. Conversely, to put work on a smaller capacity press will result in the component having to be manufactured uneconomically.

Besides taking the size and type of press into consideration for a particular job, one also has to bear in mind whether there are any other limiting factors. These other limiting factors would be laughable if only one was not totally aware of the facts. It has been known for power presses to be situated between buttresses, making it impossible to do any operation on a component over a metre in length. This form of happening does and will occur, not only in old buildings, but also in new layouts where no consideration has been given to the plant layout. This is one reason why an estimator should be totally conversant with all plant layouts within his own domain, where he can be in touch with any alteration.

Power and hand presses rely upon the accuracy of the tools that are placed in them. This is where the liaison between the toolroom, tool designer and the estimator is important. Poor tools mean scrappy work. Today many companies are finding it more economical to make tools as universal as possible, and design tools to fit standard size die sets. The handling factor within the press operation is an important one (as it is in most other operations); it is a factor which at times is overlooked. It must be remembered that an operator can keep up reasonably high speeds on light repetitive work, but when individual weights start to rise above 0.2 kg at a time, it becomes necessary to start to calculate the total weight of material to be moved within a given hour for each hour of the day. I have known that on one operation the speed the operator would have had to work would have meant that he had to move at least 2 tonnes of material every hour. Besides having to move this amount of material, there was insufficient space to have stacked the quantities required either side of the press. Upon a more realistic evaluation of the operation it was found that a quarter of a tonne was the most which could have been done in the time. This is therefore an extremely valid and viable point of view, upon which an estimator can base any argument: what can and what cannot be done. That the weight factor of material being moved through an operation is not a practical proposition, particularly for an operator over a sustained period of time; this is one point of view which management can appreciate. Again the weight factor also depends upon whether a company employs male or female labour. Another factor which effects the handling of the component is the awkwardness of the component. Some components are positively awkward to place into a tool ready for an operation, purely because of the fact that a component is of an awkward shape or length. Allowances therefore have to be made for an operator to be able to do any operation. To differentiate between operations will of course all come with experience in estimating.

### Hand Press Element

The same elements apply for hand presses as for power presses. The element is the same, although the work that is done on hand presses is of a lighter nature. Speeds that can be obtained upon hand presses, particularly by skilled workers (more so if cyclone units are used), make the hand press extremely economical with its lower overheads and low tool costs.

### Brake Press Element

Brake presses do not normally become an operation within an estimate, as brake press operations are a form of bending or cold forming and as such are dealt with as an entity on a separate estimate. I have therefore dealt with this as a separate item more closely at the end of this chapter.

## Fluid or Compressed Air Cylinder Elements

The fluid or compressed air cylinder element is either the impact operating time as a single operation or as at times where the individual impact times of a bank of cylinders are offset, the total impact times including the offset period, plus the handling time of picking the component pieces up, placing them into a jig, and taking the whole assembly out of the jig ready for the next operational cycle. During the past decade it has been an increasing practice to use fluid or compressed air operated equipment, particularly for sub-assemblies, where for instance the accurate placing of holes in relation to each other is of extreme importance, and where holes are upon different planes and angles (these types of assemblies are widely used in the motor car industry). The use of this form of equipment is now of course becoming highly specialised. The flexibility and scope of the fluid power equipment is considerable. Fortunately for the estimator and designing staffs, the major companies who produce this type of equipment not only supply the equipment, but also run suitable training courses for the various types of engineer. They also follow this up with a force of highly trained technical representatives; no estimator therefore should want for any specialised information in this field. Not only can these impact cylinders be set up in a form suitable for sub-assemblies, they can also be set up in banks, in order to pierce holes at equal or unequal distances apart, without having to resort to long bedded power presses. All that is required is lightweight jigs with the cylinders set wherever they are required. Should a greater number of cylinders be required than the power supplies can cope with, the timing of the action of the cylinders can be offset, so that there is only one handling period and the shortest of operating times between the offset.

## Bending Element

Bending can be carried out by numerous methods, as a simple press operation, as a cold rolled operation, on a brake press, jenny rolls, bending rolls (or forming rolls) or by a bending machine.

A general understanding of the term bending is any form of work carried out upon material whether by a blow or by exertions of pressure at a given point so that the material is bent or formed. Although in some industries to avoid confusion bending is carried out upon a formed section (an example of this is in the field of cold rolling of sections), strip is first cold rolled into a section profile, after which, when the straight lengths of section are manipulated to take on a shape, only then does it become termed a bending operation. A bending element can be the time of a simple press operation, or the length of time that it takes a section to be fed through a set of rolls whilst it is being swept to a radius, or being bent around a frame bending machine, plug machine, power bender or hand bender. All these have a highly specialised form of estimating expertise, which must be backed by the technical ability of the toolroom and design personnel.

### The Welding Element

Welding is one of the principle methods of joining materials together. There are, however, many different processes, the most widely known processes being:

1. Electric arc welding.
2. Resistance welding.
3. Self-adjusting arc process.
4. Inert-gas shielded arc process.
5. Oxy-acetylene welding.
6. Shielded arc process (or metal arc welding).
7. Brazing or soldering.
8. Friction welding.
9. Electron beam welding.
10. Lasers (as yet not commercial).

The welding element in practically all methods of welding is the length of time it actually takes to do a weld, plus the handling time and the cost of the materials used in the welding process for a complete welding operation. When using spot resistance welding, resistance stud welding and friction welding processes the element becomes a full operational cycle.

Most welding operations require thought to be given to the following points:

1. The accessibility of the joint to be welded.
2. The position for welding.

3. The type of material to be used.
4. The size of the electrode to be used.
5. The length of the weld (or size).
6. Whether there has to be any preliminary tack welding or not.
7. The number of weld runs.
8. Whether the weld has to be descaled, cleaned or ground flush.
9. Whether the weld has to be inspected, when certificates may have to be issued.

All these are considerations to be taken into account before being able to work out the cost of welding.

All other forms of engineering carried out by the use of machines, such as planing, lathe work, milling, drilling, etc., are elements which are determined by the individual speed and working of the machine, to the depth and number of cuts a machine can do in a given time, plus handling and setting up, all according to the type of material being used, i.e. brass being softer than mild steel can be cut easier.

Estimating elements therefore are logical events. They follow the type of work that is to be done, the type of machine or equipment being used, and in so doing become set for a particular operation. The way in which an element is measured is predetermined, and the estimator can use the same basic elements time after time. The only thing that will differ will be the actual length of time for the element.

## The Brake Press

Bending or cold forming on a brake press is a method of manipulating small quantities of short lengths of standard angles, channels and zed sections, where standard tools can be used without any additional capital cost. An important factor in the use of the brake press is to ensure that the programme for the press is geared to make the fullest use of a singular tool for over as long a period of time as possible, in order to effect the maximum usage for a single setting up operation; by doing this the most economical use is made of the tool. On a brake press an angle is formed by a single operation, a channel by two operations, as also zed sections, unless of course they happen to be lipped zed sections, when four operations will be required. More complex sections can be formed, but these are dependent upon the availability of tools, whether new tools are an economic proposition or not, or alternatively upon the experience of the brake press operator. These more complex sections also require the use of a template to ensure an accuracy of profile, as in these operations extreme care should be taken not to overbend the section. Any section which has been subjected to overbending can be difficult to rectify and may possibly have to be scrapped. It is therefore easier to form the bends as a gradual process, working towards the section profile desired, than to have to rectify them later; particular care should be taken in relation to the material springback, in order that any expensive rectification can be avoided. All these sections carry a much higher ratio of tolerance than sections formed by cold rolling owing to the method of manufacture.

The brake press operator should be a skilled man, the more experienced the better; he should also have trained assistants to help him, not merely labourers. The estimator by consulting the brake press operator can obtain extremely valuable opinions as to whether a section can be formed or not. Brake presses are invaluable for the formation of wide width sections, sections which are too wide to be formed by the cold rolling process, any limitations being controlled by the distance between the pillars of the press, the operating height and whether in fact the section can be handled. A normal profile tolerance to be set against brake press work is plus or minus 1.5mm, which of course is variable owing to the nature and complexity of the section to be formed. When estimating for brake press work, the first thing to do after ascertaining that the section in fact can be made, is to work out the amount of material required. As it is normal to cut stock steel sheet to size, it is essential to keep scrap to an absolute minimum. Having worked out the size of the material required, we set this down in the appropriate section of the estimate and work out the material price. At this stage we can also put on an allowance for any reasonable foreseeable production scrap. For example, if the section is rather complicated and we can foresee that there is going to be a heavy failure rate, possibly one in three, we must make allowances for this.

Also at this stage, or at some stage of production, if we can see that there could be some form of distortion taking place, we must allow for rectification of the distortion as a definite operation. The quantity off is of course extremely important. It may simply be a few angles that we can set in with our normal production, or it may be a specific section which needs a tool of its own, when the important question of setting up time comes into being. Setting up time is the time it takes a skilled operator and his assistant, or where there is not a skilled operator the engineering department personnel, to dismantle the tool in the press, replace it with the relevant tool, and align them so as to be able to form an accurate profile shape. This setting up time can take any length of time up to 2½ hours or more, according to the size of the press tools to be aligned. As there is no production time during this period we have the setting up time worked into the estimate in order to recover the costs. To make sure of the most economical use of tools it is important to refer to the loading schedules; also to have an appreciation of the normal flow of work through the brake press section. This is in order to be able to balance the setting up time against as great a number of orders as possible. We must also consider the operating speed of the brake press in relation to the work on hand: the brake press speed can be controlled according to the section being formed, and the way in which the pressure has to be applied to the material. In addition there is the size and weight of the material to take into consideration, and the varying number of operations per length of section. The number of operations completed can be as low as ten or twenty operations per hour; alternatively it may be a small easy angle when more than one at a time can be formed. The number could then rise to over 240 operations per hour. Also the weight and size will determine the amount of labour involved; should there be more than one assistant or not? Owing to the safety regulations within this country it is not a lawful practice for a single operator to be using the brake press, without adequate precautions, or safety guards. When safety guards are used they can be unwieldy and at times appear to be more dangerous than without them. Although weight is a critical factor in relation to the amount of labour involved, it is also critical in a reverse situation, when a large section is being formed from thin gauge material, when owing to the large surface area in relation to its thickness and weight, material will undulate. Additional labour is then required to control the material, although the weight factor does not in itself justify it. Access around the brake press and the size of the material being used has the same limiting factors as those that apply to work on all machines, whether power press, drilling machines, welding machines, etc. In addition, where very heavy sections are being formed, it is sometimes necessary to have the assistance of an overhead crane to get the work onto the brake press. This can be at times highly dangerous if the material is not correctly secured after getting it to the press, as the brake press should be actuated without use of the crane;

should reliance be made upon a crane holding the work in position, there is always the danger of the work slipping and someone being injured. It is therefore more desirable to have adequate worktables upon which the crane can present the work before it goes into the press, and for sufficient labour to be available to handle the situation. This is a field in which the estimator will have to apply his own personal experience so that there is an adequate coverage for the labour involved. It should also be borne in mind that too much labour can become a hindrance not a help. One has therefore to weigh up the pros and cons. Another factor which the estimator should also consider is the fact that too much continuous heavy work will slow down a team in a short space of time; sufficient labour must be allocated to the work efficiently; this would also have the additional safety factor that none of the labour force can come to any physical harm.

Special tools can be made for the brake press, but these can be too costly for the work involved. Tools are therefore normally only considered when it can be envisaged that there is the possibility of continuing work which could come to the company in the future if the tools were made, making it an economical proposition. At times the brake press department is inundated with work and a considerable backlog of orders builds up, making it necessary to give extremely extended delivery dates. Under these conditions I have known the sales department to give a highly inflated price rather than an extended delivery date, in the hope of losing the order. This does not always work, as it is also very probable that all of one's competitors are in the same position. If this is the case one gets the order at the high price with the acceptance of the shorter delivery date. This short delivery date has to be met, as the customer holds one to his promise. If the customer is a regular one it can also have other repercussions if it is not met. In these cases it would be far better to stick to the truthful picture and state the correct delivery date.

All work on brake presses is based upon the individual merit of each job, unless of course it can be fitted in with similar or standard work being performed. A primary function in estimating for work on the brake press, is first to eliminate as much as possible the setting up time, second to be as economical with material as possible, and finally to ensure that the most efficient number of operations and labour is being used in making the section. In doing this the estimator must make himself well acquainted with the present and future work load of the brake press section.

## Formulae for Air Bends Upon a Brake Press

Table 1

| Metal thickness (mm) | 1.0 | 1.2 | 1.6 | 2.0 | 2.5 | 3.0 | 4.5 | 6.0 |
|---|---|---|---|---|---|---|---|---|
| Factor $F$ | 2 | 2½ | 3½ | 4¼ | 5½ | 6¾ | 10 | 14 |

The load to bend any thickness and length of plate under air bend conditions to a right angle can be calculated from the following formula:*

$$P = \frac{8 \times F \times L \times T}{W \times Mf}$$

where $P$ = lead to bend in tonnes,

$F$ = factor from Table 1 above,

$L$ = length of metal to be bent in mm,

$T$ = thickness of metal in mm,

$Mf$ = Material factor from Table 2 below.

Table 2

| Material | Factor |
|---|---|
| Mild steel | 1 |
| Soft aluminium | 2 |
| Aluminium alloys | 1 |
| Soft brass | 2 |
| Stainless steel | 2/3 |

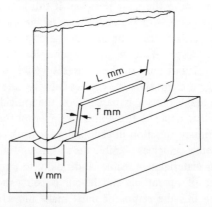

*The tables and formulae for air bends are reprinted by kind permission of The Bronx Engineering Co. Ltd., Lye, West Midlands.

The loads of which the brake press or presses are capable of achieving within the estimator's own company should already have been tabulated, in sizes, as well as all pertinent details such as the distance between pillars, size of bed, etc. It is also extremely useful to keep a set of plastic cut-outs of the form of all brake press tools in the company's possession, with the length of the tool marked upon it, as this is far more useful than a mere drawing or sketch of the tool. The plastic profile of the tool can be set against any drawing (full size) to check whether the tool is the correct one for the job or not.

# CHAPTER 5

# WORK MEASUREMENT

Work measurement is a technique used by the work study department, and the results of any work measurement should be made available to the estimating department, for their evaluation and use. The results of any evaluation of this technique are normally used for the following purposes:

1. Production control and planning.
2. Cost control.
3. Plant and equipment utilisation.
4. Budgetary control.
5. Comparison of methods and techniques.
6. Financial incentive schemes.

Any evaluation of work measurement should also be used in the estimating department, as it is essential that the estimator not only understands work measurement and work study methods, but is also able to utilise the evaluation of these methods and techniques within the estimating function to enable him to reach a greater degree of accuracy and efficiency.

Here there is a fundamental difference which is not clearly understood between the functions of the estimating department and the work study department. Whilst the work study department is timing and measuring actual occurrences, the estimating department is measuring the hypothetical, what is likely to happen, not what is happening. There are also differences of opinion regarding this distinction and the meaning given to the definition of estimating and its functions, particularly by work study personnel.

There should be a good feedback of information from the work study department to the estimating department, because of this differentiation between them. In this there is no reason why estimates should not reach a high degree of accuracy for the timing of operations upon a hypothetical basis, provided that both the estimator and work study have evaluated previous recorded work correctly. It is unfortunate that the measurement of work is not well received by workers on the shop floor, as it is necessary to have some form of work measurement, both as a basis for future estimating and for incentive schemes. Again it is unfortunate that it has not been possible to have discovered a perfect incentive scheme. Each and everyone has some drawback, and it is indeed fortunate that the estimating department has a buffer between them and the workers in the shape of the work study department.

The same stigma does not attach itself to the estimating department. Rather there seems to be more sympathy for the department, stemming from an appreciation of the difficulties which confront the estimator. (Workers somehow appear to appreciate estimating difficulties far greater than management.) The standards for all work measurement or time study can be found in British Standard Specification 3138/59/31001, where the definition of time study is: "A work measurement technique for recording the times and rates of working for elements of a specific job carried out under specified conditions and analysing the data in order to obtain the time necessary for carrying out the job at a defined level of performance." In work measurement there are definite and ordered sequences to follow in procedure, i.e.:

1. Selecting the work to be studied.
2. Recording the facts of the work to be done, and the methods and elements of activity.
3. Measure each element in terms of time over a sufficient number of cycles, in order to ensure a representative picture.

In estimating, the above is not possible as one is working upon the hypothetical:

1. One cannot select the work to be studied. All that one can do, based upon past experience, is to select an area of work or an element of work.
2. One cannot record any facts of work as none have been done. Again one can only by the study of historical data and personal knowledge split the work up into the actual elements which one thinks will apply.
3. One cannot measure time for doing the job; one can only assume by training and experience that the work can be done in a particular time, and by doing so build up a complete picture of how long it will take to do a singular piece of work.

The added procedures of examining recorded data and element times, compiling a time for the operation and eventually defining a precise series of activities and a standard time for these, are luxuries that the estimator is unable to employ. Having decided upon how long it will take to do a particular element within an estimate, the estimator is stuck with it. There is, fortunately, the fact that as an estimator builds up the total cost, he is able to tell by experience whether or not the final price is emerging within the right area. Particular estimators, through their experience and total knowledge of their industry, can define within reasonable limits the amount of an estimate for particular forms of work which fall as a regular item, that is with a definite percentage of labour, material or overhead requirements. An analysis of comparative work will find that each of the labour, material and overhead elements of an estimate will fall as a certain percentage of the total cost of the item. Any marked deviation from these percentages will denote an element of error within the estimate. This is one of the more positive checks that an experienced estimator can make upon his estimate.

## Standard Time

Standard time is derived from the combined evaluated times of the observed time plus rating allowance plus relaxation allowance plus contingency allowance, where the observed time is the time that has been measured by a trained time study man, to which a rating allowance has been added according to the assessment made at the time of the worker's speed and ability. Relaxation allowances are predetermined allowances for personal needs and fatigue, applicable to the job on hand, to which is added a contingency allowance to cover any minor or infrequent elements which cannot be justified. It is useless to have any standard times unless they have been correctly worked out and are acceptable in written form to the necessary involved personnel.

Standard time

| Basic time | R. A. | C. A. |
|---|---|---|

Basic time = observed time + rating allowance.

R.A. (relaxation allowance) = personal needs + fatigue allowance.

C.A. (contingency allowance) = any minor or infrequent elements that cannot be justified.

From this it will be seen that the work study department relies in the first instance upon accurate observed ratings for their basic time, which also includes an assessment of the individual's capability for the work done.

This individual ability or rating allowance is an element that the estimator cannot use. He cannot rate any worker different from another, as he, at the time, cannot know who is going to do the work. It is essential for the estimator to use only what can be described as average ratings — what a reasonable man can expect to be done within a period of time for a particular piece of work, based upon any historical data or personal experience. Relaxation and contingency allowances are treated differently. They are not so clinically analysed, as within work study; it is more from personal judgement and assessment of work, because it is done within the estimator's own factory. This determines these allowances, as these allowances are based upon localised conditions. Work measurement is best related to work studies standard times, but because of various elementary differences in the manner in which work is approached, it will be found that the estimator should relate the number of actual operations to a given time. The number of operations to a given time also makes it simpler for any calculations to be made for the apportionment of labour and overhead costs. These operational unit times will be determined by the various industries, but in all they will follow a basic pattern, in which the estimator should be trained.

# CHAPTER 6

## ESTIMATING FOR MASS PRODUCTION AS EXPERIENCED IN THE AUTOMOBILE INDUSTRY

The automobile manufacturers of today are only interested in the final assembly of the motor car on the track. The actual production of components is carried out on sub-contractor's premises, and they are subjected to inspection either at the sub-contractor's or the receiving depot. It is also extremely prevalent in the trade that the automobile manufacturer has more than one sub-contractor for each component, unless the components are of a special nature, or a small number off; this ensures continuity of supplies. Should a single sub-contractor have any difficulties, other sub-contractors can make up the quota, and there is then no stoppage on the assembly lines. Stocks held by the automobile manufacturer do not normally last for more than a few days. It is the sub-contractors who act as stockists, delivering to a tight schedule. This method has the effect that should there be any major stoppage on the assembly lines, an accumulation of supplies builds up at all sub-contractors. Although this system of multiple suppliers ensures the continuity of supplies, it does not necessarily mean that the competition for the major part of the orders keeps the sub-contractors on their toes and components up to specification. In fact it can have the reverse effect, owing to cut-throat competition. Some suppliers get away with manufacturing the components from the thinnest material possible, due to the fact that mill tolerance upon thin gauge material is plus or minus one gauge. This results in the certainty in many cases of material being one gauge light. The automobile manufacturer might specify 0.9mm to 0.7mm permitted thickness, and the sub-contractor orders 0.7mm, anticipating receiving it with a plus tolerance of up to 0.8mm. In actual fact he can receive anything down to 0.6mm, and it has been known for components to be manufactured from the thinner material, only to be rejected by the manufacturer, who has accepted them at a later stage when shortages would have stopped production on the track. Stoppages of the assembly

44

line, particularly if they should be caused by strikes, have the effect of laying off workers in the sub-contractors if they do not have any alternative forms of work to fall back upon. Sub-contractors are unable to stockpile any large amount of components, as they are subject to modification at any time. Also considerable sums of capital can be tied up in materials and storage space, explaining why a strike within the automobile industry has such a widespread effect almost immediately. Sub-contractors, therefore, plan their production to coincide with production schedules as laid down by automobile manufacturers, tying statutory holidays, etc., in with those of the automobile industry and only stockpiling components to cover any slight differences in these periods, or to allow for inclement weather which could delay the supply of components to a manufacturer. Components are also scheduled for different models in production and it can be possible for one component to be used on a number of models. This, in turn, has an effect on the delivery schedule, some components being large in numbers, some small, particularly where there are right- and left-hand components for overseas models. The quantity off per component and the scheduling of components is therefore extremely important and all the various aspects which this entails have to be taken into account when preparing an estimate.

Another major consideration where cold rolled sections are used is the section profile. Not only can it be common to many components but also to components of different manufacturers. In these cases it becomes more economical to do a long production run of the section, stocking the section rather than rolling for each different order. Automobile manufacturers in the past few years have also become more cost-conscious, particularly when putting forward a new model, as the cost of tooling up, even for a few decorative strips or flashings for one model, can run into many thousands of pounds. Wherever it is possible to standardise upon sections and materials, manufacturers do so, as they have only the body line tools, jigs and formers to pay for, saving the heavy cost of rolls. Manufacturers in the design of cars do attempt to retain as many of the basic formations as possible in order to cut the total capital expenditure of a new model down to a feasible level. Sometimes it is possible to have existing tools modified, but here there is a snag. These may be required to supply replacement components for an existing model or one particular model which is being kept in production. This is where the fact that there are three component sub-contractors comes into its true significance. You stop supplies from two, but retain the supplies on a lower level from the third.

The question of replacement components also causes problems for the automobile manufacturer, particularly when tools have been used for an extremely successful model and are worn out. What expense should be incurred in order to be able to supply a limited number of components for replacements?

Sometimes it will be better to make them on worn-out old tools and rectify them, or even make them on a prototype basis, rather than to go to the expense of further new tools for a limited life. What is a reasonable length of time for a component manufacturer to keep tools, 5 years, 10 years, 20 years after ceasing to manufacture for production models? Replacement components have to be made to cover all the repairs that are made to vehicles, whether from accident or old age. Every enquiry has its many facets and each and every one will have to be dealt with according to its merits and the situation prevailing at the time.

Before we can estimate, we must have an enquiry. For our example let us assume that we have received an enquiry from an automobile manufacturer for components as shown in Fig. 2 and the accompanying specification and drawings (Figs. 3a, 3b, and 4), the enquiry having been passed to the estimating department by the sales department, first having been registered and acknowledged. From the enquiry we find that we have to quote for pairs of a motor car window channel, to be supplied at the rate of 150 pairs per week for the first four weeks, stepping up to 850 pairs per week for the following two weeks, and from then onwards at a weekly rate of 1500 pairs for full production. The price is to be firm for a period of six months and the components to be formed as per drawing AYE 3785421/2A, all to be in a standard matt black finish as per specification B627. The type of component shown in the drawing is one which follows a general form for this type of component within the automobile industry. It is a channel that extends from the window opening downwards inside the body of the door, and which keeps the glass of the window in position as it is raised or lowered. Before it is inserted into a vehicle it has a felt insert fixed within the channel in order to protect the glass, allowing it to slide up and down freely. Also for the benefit of this exercise we are treating it as a new concept and design. Having received the enquiry, we ensure that we have all the relevant data, etc., in order to enable us to make a correct estimate. We examine the

THE AUTOMOTIVE ENGINELESS AUTOMOBILE COMPANY LTD

SPROGGINS LANE:  CARFUL. CA1 6FD:  SKYE

Telex:  69784

Telephone:  SKYE 35693

10th August 1975

CJRA/BC

The Windmore Engineering Co Ltd,
Engine Lane,
Muckleborough,
BRUMHAN.

Dear Sirs,

       Please quote your keenest price for the following,
Car window channel right hand, drawing number AYE 3785421/ A
Car window channel left hand, drawing number AYE 3785422/ A
to be delivered at the following scheduled rate:

> 4 weeks period 11 1975, 150 pairs per week.
> weeks 1 and 2 period 12 1975,  850 pairs per week.
> week 3 period 12 until end of period 6 1976, at the
> rate of 1,500 pairs per week.

       Price to be given per component and to be firm for the
period of contract, all tooling to be shown as a separate item, components
to be finished as per specification B 627.

       Quotation must be received by the 10th September 1975 in order
to be considered.

Yours faithfully,

*J. Bloggs*

Chief Buyer.

**Fig. 2.**

drawings, find that it is clear enough and that the enquiry itself is definite
and understandable. From the drawing we find that the component
required is basically three different pieces welded together, the main
section being a cold rolled section, in a straight length clipped out at one
end, with piercings along its length and two small pressings spot welded to
it, one at each end. The larger of the pressings has a "D"-shaped clinch nut

fixed into it, the whole, after manufacture, being painted black as per specification. The material for the cold rolled section is 1.0mm and that for the pressings 0.8mm for the smaller of the two pressings and 1.2mm for the larger pressing, material which we carry in stock. The "D"-shaped clinch nut, however, is a special and subsequently is not carried in stock.

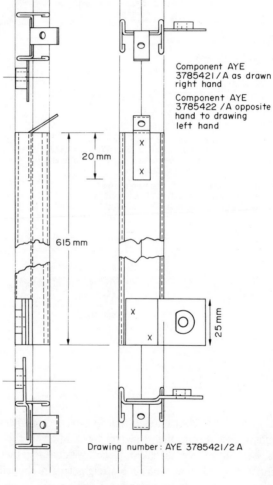

Component AYE
3785421/A as drawn
right hand

Component AYE
3785422/A opposite
hand to drawing
left hand

20 mm

615 mm

25 mm

Drawing number: AYE 3785421/2 A

Fig. 3a.

In this case it is important that an enquiry for the clinch nut is sent off as quickly as possible in order to obtain a price and delivery. The enquiry is made out for the whole of the six months' supply of nuts by the purchasing office upon receipt of instructions from the estimating department. We can do this with relative safety as if there should be any changes in design it is most likely that it will effect the nut, therefore a keener price can be obtained for the larger quantity. Having sent off the enquiry via the purchasing office, our next concern is whether we are able to manufacture the cold rolled section as it is drawn. If we can, what

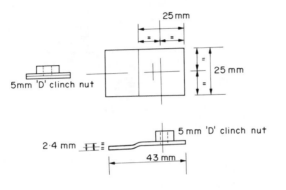

Material : 1·2 mm

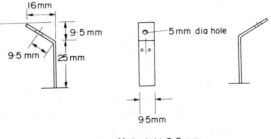

Material : 0·8 mm

Fig. 3b.

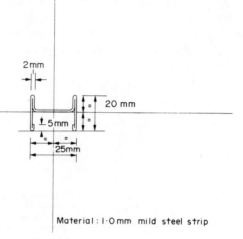

Section   profile
Components   AYE  3785421/2A

2mm

20 mm

5mm

25mm

Material : l·0 mm  mild steel strip

Fig. 3c.

number of rolls will be required to roll the section and upon what machine
will it be rolled? (To us it is a new section.) It must be remembered that
the design of cold rolled section rolls is a highly skilled job and the
estimator will have to obtain expert advice. For this it will be necessary to
consult the roll designer who will let us know the answer to these
questions. The cold rolling process is a process whereby light gauge
sections can be formed by a continuous operation from coiled strip. The
strip passes through a series of rolls, all of which are shaped to a varying
degree of the section profile until the section is finally formed through the
last pair of rolls, after which it may or may not pass through a further set
of sizing rolls, depending upon the accuracy required. Sections can be
formed from various types of metals. Forming rolls are set upon either an
inboard rolling machine or an outboard machine. An inboard rolling
machine, as the name suggests, is where the forming rolls are set between a
pair of housings to give greater strength and power to the rolls, and upon

THE AUTOMOTIVE ENGINELESS AUTOMOBILE COMPANY LTD

<u>Paint Specification B627</u>

1.   <u>General description</u>

The finished paint film shall be a single coat of matt black and shall form a suitably smooth and uniformly, firmly, adherent and serviceable protective coating for miscellaneous steel and ion chassis parts.

2.   <u>Finished paint film characteristics at 20 – 25° C</u>.

   (i)    <u>Thickness.</u> Not less than 0.012 mm

   (ii)   <u>Bend test.</u> 180° bend over 6mm Radius without damage or detaching.

   (iii)  <u>Adhesive and Flexibility</u>. Indentation test of 2mm without initial cracking; indentation test of 3mm without flaking off.

   (iv)   <u>Hardness</u>. As determined by D. T. D. scratch test – 750 gms min.

   (v)    <u>Resistance to petrol and oil.</u> Immersion in petrol for 1 hour, or in a mineral oil for 2 hours at 20° C. shall not develop any signs of permanent injury.

   (vi)   <u>Corrosion test.</u> Shall withstand 48 hours salt spray test (20% solution) without any blistering or flaking, or rusting of bare metal or serious rust spread from bare edges.

3.   <u>The following paint materials</u> or officially approved equivalents can be used.

DY 127/9 Enamel black matt undercoat (dip and bake)
XY 56/13 Enamel black matt single coat (dip and bake).

Signed.

*P. Green*

Chief Chemist.

Fig. 4.

which the heavier sections are rolled. The outboard, on the other hand, has the rolls set upon spindles on one side of a housing. Sometimes they are also set upon both sides of the housings. These are used for the smaller lightweight sections, and it is possible to roll two sections at a time, one either side.

Figure 5 gives an illustration of the two types of machine.

The cold rolled method of forming sections is a relatively cheap method where a considerable footage is required of the section, particularly when

Rolling machines

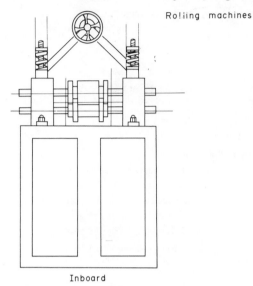

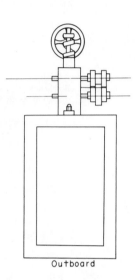

Inboard

Outboard

Fig. 5.

long lengths of the section are required. In normal production runs, the general practice is first to cut the section to approximately 30,000mm lengths, then to cut it to shorter lengths afterwards. When cutting to shorter lengths it is possible to carry out further operations such as clipping the ends or piercing the section. It is also possible to carry out some operations to the strip prior to rolling. A normal delivery of random section lengths to customers is between 4500mm and 6000mm rolled at speeds which varies from between 182,000mm per hour to 1,220,000mm per hour and in the case of some extremely light and small sections even faster. If we now take a look at Fig. 6 we will see the progression of stages for the section we are estimating. It requires six stages, each stage consisting of a pair of rolls, until the final form takes shape. In this example we shall assume that a tight tolerance is required, which will make it necessary to have an additional pair of sizing rolls. Each pair of rolls is made up of a bottom and a top roll and therefore to the initiate Fig. 6 may take a little studying. Having now, through the aid of the roll designer, decided upon the number of rolls required, the designer will also let us know what the size of the rolls is going to be and the size of the

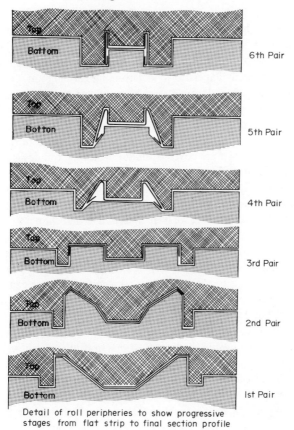

Detail of roll peripheries to show progressive
stages from flat strip to final section profile

Fig. 6.

machine upon which to roll the section. For our example it has been
decided that a medium-sized machine will be adequate for the rolling of
the section. This can be done at a speed of approximately 550,000mm per
hour. All enquiries that are received which require new rolls are passed on
to the roll designer, so that he can decide whether the section can be
formed; he will then return the enquiries back to the estimating
department with all the relevant information. In most cases the designer
will require time in which to come to a correct conclusion. Arriving at a
wrong conclusion at this stage could result in the loss of not only capital

but of orders. Owing to the great variety of new section profiles, all of which have to be machined from roll blanks, prices cannot be pre-determined by a system of set charges; the cost of rolls are therefore fixed for a basic size of roll 'and an average form of section, upon which the designer will use his expertise in determining the amount of additional machining required and apply a system of percentages which are set against the roll costs in order to determine what the final cost will be. This system of percentages can uplift the machining costs of rolls from anything like 10% to 80% above the cost of an average set of rolls. By this system tool room costs are covered and it enables the roll designer to cover costs for any profile shape.

Once having decided upon the number of rolls and the size of the rolling machine, we are now able to start estimating for the cost of rolling the section. For our exercise we have determined that six pairs of rolls plus a pair of sizing rolls will be required. The section will be rolled upon a medium machine, and as the section is reasonably straightforward we are having no uplift percentage for any additional machining costs. There is, however, in addition to the roll costs a trying-out cost. One cannot expect rolls to perform perfectly the first time that they are put onto a machine for rolling. They have to be tried out in conjunction with each other, and there may have to be further machining before they are perfect. Besides the checking of the profile shape, the estimated width of strip must also be checked for the correct section profile. When costing the rolls, we find from our list of roll charges that a pair of rolls for a medium machine is £165.00 per pair, and the cost of sizing rolls £114.00 per pair, with the cost of trying out the rolls set at £45.00 per pair. With no further uplift we are able to set out the cost of the rolls for our section. This would be as follows:

|  | £ |
|---|---|
| 6 pairs of medium rolls at £165.00 per pair | 990.00 |
| 1 pair of sizing rolls at £114.00 per pair | 114.00 |
| 6 pairs of trying out rolls charges at £45.00 per pair | 270.00 |
| Total roll costs | £1374.00 |

Having worked out the cost of rolls, we next need the cost of the strip, and the normal practice is to work out the width of strip mathematically along its mean axis, or to step around the section profile using a pair of dividers in 1.0mm steps, by which method the section profile can be estimated to within 0.25mm. This is accurate enough for mild steel as the cost of 1.0mm of 1.0mm mild steel strip, 100mm long, for instance, would be less than 1/100th of a new penny, which for all intents and purposes is of little consequence to be more accurate, and for our section we can take it for all practical purposes to be 77mm wide. In practice the width of section is checked by the trying out of rolls procedure, as with the bends of any section the stresses and strains which take place in the cold rolling process will either stretch or compress material, which will in itself also make a difference to the actual width of strip. Should every width of strip be analysed for these stresses and strains and each and every bend carefully worked out to find out these implications in the rolling of the section (a long and tedious task) it would be possible to estimate to an extreme accuracy, but the time alone in doing so would make it most uneconomical. Having ascertained that the width of strip for the section profile will be 77mm and that the gauge of material required is 1.0mm in mild steel, we find that the mill cost for 30 to 70 tonne lots of 1000mm x 1.0mm thick coiled strip is for our exercise £150.00 per tonne; it is from 1000mm wide coils that we would shear our 77mm width strip, doing so upon a rotary gang slitter. The cost of shearing the strip on a slitter would be £6.00 per hour for labour and £15.00 per hour for overheads, an average coil being 6 tonnes in weight; 1.0mm thick material being approx. 6kg per 1000mm$^2$. A coil would, therefore, be approximately 1,000,000mm of 1000mm wide strip, from which we would obtain 12,000,000mm of 77mm strip. Through our rotary gang slitter we can safely say that we will pass one coil of 1000mm wide material per hour, which would be 12,000,000mm of 77mm strip for £21.00. This would make a cost of 0.18 new pence per 1000mm of strip. Having the cost of slitting the strip and knowing the cost of the material, i.e. 12,000,000mm (6 tonnes), 6 x £150.00 = £900.00, therefore 1000mm of strip costs 7.50 new pence. In addition to knowing the cost of the material we have to ascertain the amount of scrap used in the trying out stage. This, of course, is a variance, depending upon the complexity of the section profile, and therefore can vary from 3000mm to 6000mm of material or even more. However, to overcome this variance it is a normal practice to put a 2½% scrap value on the strip for rollings of 6,000,000mm or more, as in practice this percentage has been found out by the accountants to be sufficient to cover the cost of scrap in the majority of cases. A larger value, however, is placed upon shorter rolling quantities.

As previously stated, we have decided to roll the section upon a medium-sized machine, which will roll strip at an average speed of 550,000mm per hour. The machine overheads are £75.00 per hour and the labour costs £12.75 per hour (£87.75 for 550,000mm). This works out at 15.90 new pence per 1000mm for labour and overheads. There is, however, the additional cost of setting up the machine, which in this case will take 4.25 hours at a labour cost of £15.00 per hour plus the overheads of the machine at £75.00 per hour, making a cost of £90.00 x 4.25 = £382.50 for a standard economical rolling of 12,000,000mm = 3.19 new pence per 1000mm. We could therefore enumerate our rolling costs as follows:

|  | new pence |
|---|---|
| Material cost | 7.50 |
| Slitting cost | 0.18 |
| Rolling cost | 15.90 |
| Setting up time | 3.19 |
| Trying out scrap, 2½% | 0.67 |
| Total cost per 1000mm of section | 27.44 new pence |

To this we must add a handling charge, which again is expressed as a percentage in order to recover the incidental costs of handling material from stock to the slitter on to the forming machines. This charge is also a variance, depending upon the amount of material to be shifted. For our example a simple 5% handling charge would be *sufficient* to recover these incidental costs. These variance percentages are set out by the accounts department, and in many cases estimating departments have attempted to do away with these for actual costs. However, accountants seem to prefer to have a standard percentage recovery costs, because what is lost on the swings is gained on the roundabouts, without having any complexity of working out individual incidental costs.

Therefore we have 27.44 new pence plus 5% = 28.81 new pence per 1000mm. As the component is 615mm long it will cost 17.72 new pence for the basic cold rolled section (see Fig. 7).

| 77mm x 615mm x 1.0mm thick mild steel at £150.00 per tonne. 1 coil = 6 tonnes = 12,000,000 mm of 77mm strip. | | Slitting 12,000,000 mm per hr at £21.00 per hour | 0.18 |
|---|---|---|---|
| 1000 mm = | 7.50 | Rolling on medium machine. Rolling speed 550,000 mm per hour. Overheads £75.00 labour £12.75 = £87.75 per hour. | 15.90 |
| | | Setting up machine at £15 per hour labour plus machine over heads £75.00 per hour = £90.00 per hour for 4.25 hrs = £382.50 for a standard economical rolling of 12,000,000 mm | 3.19 |

|  | Material | 7.50 |
|---|---|---|
|  | Total | 26.77 |
| 2½% trying out scrap | | 0.67 |
| Cost per 1000 mm of section | | 27.44 |

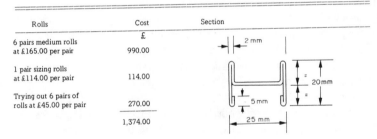

| Rolls | Cost £ |
|---|---|
| 6 pairs medium rolls at £165.00 per pair | 990.00 |
| 1 pair sizing rolls at £114.00 per pair | 114.00 |
| Trying out 6 pairs of rolls at £45.00 per pair | 270.00 |
| | 1,374.00 |

Material: 1·00mm mild steel strip

Fig. 7.

Having the basic cost of the section we must now carry on and obtain the cost of manipulating the section, although this is only a simple straight section which has to be pierced and clipped. First we must cut the section off to length, an operation which is either cut off on a power press or cut off by power saw. Our section can be cut off on a power press, for which operation we will need a cutting off tool, the cost of which will be £175.00, cutting off short lengths of section at a speed of 600 lengths per hour, using two women as the labour force. Clipping out one end of the section will need another tool, for which we will require £189.00, but as there is a left-hand and a right-hand component we shall require two of these tools, the operation being carried out upon a power press by a woman operator at a speed of about 277 operations per hour. It only remains to pierce the holes along the length of the section, which can easily be done by compressed air impact cylinders. This, however, will depend upon whether or not it is the practice of the company to use compressed air or not. We will assume that it is and that we have the necessary cylinders. This, in turn, cuts down the cost of tooling, as we only have to make a simple jig for the operation. Even if we do not have the correct size of punches and dies, the cost would be a moderate £75.00, the cylinders being set in line along the jig. The component is laid in the jig and clamped into position, the cylinders actuated and the piercing completed. All this is capable of being done by a woman, at a speed of 150 operations per hour. In Fig. 8 we can see how to set out the cost of materials and the operations and also how to write down the labour costs and overheads, thereby working out the cost of the component. We still have to work out the cost of the two pressings required to complete the component, both of which can be manufactured from coiled strip, cutting off the length of material, flattening the material and in the case of the smaller of the two pressings piercing the small hole at the same time.

This would require a slide feed tool costing £135.00 which would be capable of 3500 operations per hour, one man looking after the press. This is because of the need to feed the machine with coils of material too heavy for a woman to handle. Where there are several slide feed presses, a woman can look after the machines, whilst a man could feed several machines with material. We will, however, for our example and the insignificance of the labour content upon this operation assume that the labour force is one man. The small pressing therefore only has to be cranked to be completed, an operation which can be easily carried out upon a hand press, particularly if a "cyclone unit" has been fitted, enabling an operator to achieve speeds in excess of 1000 operations per hour, the cost of a tool being £35.00. We are now left with the larger of the two pressings, for which there are several alternative methods of manufacture. It would be possible to have it made as a cold rolled section, cut off and pierced, made upon a multi-stage press tool, or in a similar manner to the small pressing. For this exercise, we will choose the multi-stage tool (cost £280.00), where the number of operations will be 3500 per hour and for the same reason as before we shall put one man on the operation for the labour

Customer:      The Automotive Engineless Automobile Company Ltd

Description:     Window Channel Section

Drawing / Component Number:      AYE 3785421 / 2 A

Date:  . . . . . . . . . . . . . . . .

| Material | Cost | Operation | No. off per hour | Labour | Over-heads |
|---|---|---|---|---|---|
| | p. | | | p. | p. |
| 1.0mm section as per estimate. 615mm long at 27.44p per 1000mm | 16.880 | Cut off. 2.w. 1 op(3.0) small P/P 1222 (7.50) | 600 | 0.500 | 1.248 |
| Handling 5% | 0.840 | Clip out one end 1 op | | | |
| 0.8mm thick coiled strip. 1 off 9.5mm x | | 1.w. small P/P 1222 (1.50)       (7.50) | 277 | 0.542 | 2.706 |
| 45mm = wt. 0.0024kg at £186.00 per tonne = 18.60p per kg. | 0.045 | Pierce 4 holes. 1 op 1.w. C.A.m/c 1048 | 150 | 1.000 | 4.200 |
| Handling 10% | 0.005 | (1.50)       (6.30) | | | |
| 1.2mm thick coiled strip. 1 off 25mm x 50mm = wt. 0.011 kg at £186.00 per tonne = | | Cut off flatten and pierce small pressing 1.m. 1 op. slide feed (2.00) press 1225 (9.00) | 3,500 | 0.006 | 0.258 |
| 18.60p per kg. | 0.205 | | | | |
| Handling 10% | 0.021 | Crank 1 op. 1.w. (1.50) cyclone unit 1892 (6.00) | 1,000 | 0.150 | 0.600 |
| 'D' clinch nut ex A.B.C. Co quote, 1 off at £4.11 per 100. | 4.110 | Form, pierce & cut off 1.m. 1 op. (2.00) slide feed press 1880 (9.00) | 3,500 | 0.006 | 0.258 |
| Handling 10% | 0.411 | | | | |
| | 22.507 | Clinch 'D' nut 1.w.(1.50) Hand press 1880 (5.25) | 240 | 0.624 | 2.187 |
| | | Spot weld 1st op 2 spots 1.w. (2.00) 25kva 1628 (6.75) | 135 | 1.480 | 5.001 |
| | | Spot weld 2nd op 2 spots 1.w. (2.00) 25kva 1628 (6.75) | 135 | 1.480 | 5.001 |
| | | Sub totals | | 5.988 | 21.459 |

|  |  | |
|---|---|---|
| | Total | 28.447 |
| | Material | 22.507 |
| | Paint | 5.163 |
| Overall allowance 5% | | 2.806 |
| Cost each | | 58.923p |

| Tools | Cost |
|---|---|
| | £ |
| Rolls | 1,374 |
| Cut off | 175 |
| Clipping tools | 398 |
| Piercing tools | 75 |
| Slide feed tool | 135 |
| Crank | 35 |
| Slide feed tool | 280 |
| Clinch tool | 45 |
| Weld Jig | 25 |
| Weld Jigs | 50 |
| Paint grommets· | 25 |
| Total | 2,617 |

Tool costs £2,617.00

**Fig. 8.**

content. (In this particular case the relevant cost between male and female labour would be extremely fractional, in fact 0.012 new pence per article.) All that remains is for the 'D' clinch nut to be fixed into position. This is an operation which can be done on a hand press by a woman at approximately 240 operations per hour, the tool costing about £45.00. All that now remains is to assemble all the three items into one component. This is done by spot welding each pressing upon the end of the cold rolled section in the positions shown on the drawing. To do this we shall require spot welding jigs for which it is the normal practice to ascertain the best method of inserting the individual parts, so that they fall naturally into position for welding. If it is possible for them to fall into a natural position where the components lay flat upon each other, it may be possible to dispense with the necessity of using clamps to hold the work in place, which again allows the spot welding to be done much quicker. The provision of correct jigs is important, having as it does a direct bearing upon the number of operations per hour; in all these cases jig designers should be consulted. Both of these pressings can be done in a simple manner, two spots per end, for which a fair average number of operations would be 270 spots or 135 individual ends. As one pressing is right- and left-handed we shall require three jigs for all of these operations; simple jigs of this nature can be made for £25 each. We have now completed the mechanical side of the components manufacture. All that remains to be done is painting the components to the specification stated and the conditions laid down by the specification (Fig. 4). First the component has to be degreased and granodised (a process which etches the surface of the material). Most paint shops are laid out for a continuous process, so that components are painted straight after the cleansing operations. Components are placed into mesh baskets which are lowered into each vat in turn, vats being any length according to the type of work undertaken by the company. In our case the vats and baskets are 6000mm long. Each mesh basket is termed one load, and one basket in our case will hold 600 components, at a cost per load of £12.75. After being degreased, granodised and washed free from chemicals and dried, the components are ready for painting, the painting operation being done as soon as possible after cleansing otherwise oxidisation will take place on the unprotected metal. This is where correct workload schedules are necessary to ensure a free flow of material. The chassis black finish which has been called for can be done by dipping the component in the paint, then stoving the component. The screw thread of the 'D' clinch nut must, however, be free of paint. This thread is protected by the simple expediency of having a dipping hook with a nylon grommit fixed upon it which will fit snugly into the thread of the nut and which also will be the means of hanging the component up during the stoving operation. The component has the hook fixed into position by one man who passes it onto the second operator who will immerse the component into the paint bosh (tank), after which he will hang it up upon a moving overhead track, which follows a path over a drip tray (which catches the surplus paint) into and through the infra-red stove. Operating in this manner two men will

Customer: The Automotive Engineless Automobile Company Ltd

Description: Window channel section

Drawing / Component Number: AYE 3785421/2 A        Date . . . . . . . . . . . . . . . .

| Material | Cost p. | Operation | No. off per hour | Labour p. | Over heads p. |
|---|---|---|---|---|---|
| Black paint DY 127/9 at 48p per litre. Coverage 72,000 square mm per litre = 0.0007p per square mm Area approx 490 sq mm | 0.343 | Dip and stove 2.m. (4.00) Infra red stove 3048 (11.25) | 480 | 0.832 | 2.340 |
| Handling 10% | 0.034 | | | | |
| Waste + 5% | 0.019 | | | Total | 3.172 |
| | | Degrease and granodise | | | 1.991 |
| Degrease and granodise 600 off per load at £8.70 per load | 1.450 | | Cost each for paint | | 5.163p |
| Handling 10% | 0.145 | | | | |
| | 1.991 | | | | |

| TOOLS | COST £ |
|---|---|
| Nylon grommets | 25 |
| (hooks available) | |

Fig. 9.

handle approximately 480 components per hour. When the component comes through the stove, it is taken off the track and hung up upon a stand to cool down, after which it is packed for despatch. The question of the cost of despatching components is dealt with in different ways by different companies, as where mass production methods are used for small components the cost of packing and despatch is an insignificant amount per component. Where this occurs it is usual to incorporate the costs as part of the overheads. Where the component is large enough for an individual cost to be made, this is done. For components like our example they are merely bundled into tens and hung up in special crates for the purpose and charges are recovered as part of the overheads. Having gone through all the various operations we enter them as we go upon our estimate sheet (Fig. 9) entering all the relevant costs, working them out and arriving at a price or cost of the component, including all the relevant tooling costs as a separate item. All this is passed on to the sales department who will put in the final profit margins and make out a quotation to the customer.

We have, therefore, gone through all the stages of a simple estimate for the automobile industry. From this simple example it will be seen that estimating is a system of logical steps, enquiry, deliberations, formulation, evaluation and calculation, all to arrive at an end price or value.

[*Note:* In this example the various figures should not be taken as actual costs; they are figures which have been used purely as an illustration to show how an estimate is formulated.]

# CHAPTER 7

# AIDS TO ESTIMATING

One of the most important aspects of estimating is the existence of any information or data which could be of assistance in the preparation of any estimate, all of which could have a fundamental bearing upon the accuracy of an estimator's forecast. If estimating is to be treated as an art or management technique in order to obtain orders rather than a system in order to obtain a price, then the outlook upon the subject would be considerably different, and assistance would be forthcoming as a natural progression from manufacturing and production methods and techniques. It is fundamentally an educational approach that the estimator must insist upon, to be able to show other departments the benefits that can be obtained by the retrieval of information about manufacturing and production methods, as it is from this source of information, upon which the estimator relies, that he is enabled to set up his historical data bank. All information can be fed back to the estimating department in many different ways and each company has its own methods. In many cases this information is unfortunately completely ignored by the estimating department and the estimator must not take it for granted, and expect everything to be done for him. It is only natural for any department who has been feeding back information to the estimating department upon discovering that it goes into the nearest waste paper basket, for them to cut off the offending department from their circulation list, rather than to continue to supply what is apparently unwanted data. It is the prerogative of the estimator himself to find out what feedback of data and information is available, each company having its own system. Each estimator must therefore find out where and to whom the data and information is passed. Usually it can be found that a separate feedback is unnecessary as one department is collating all the relevant data in the manner desired, in which case all that one has to do is to arrange for a copy to be made and circulated. There are, however, many occasions where too much work is duplicated, in which case this should be re-arranged and circumvented. The first stages of the feedback of any information and data will naturally come from the shop floor. There are in these cases formal documents which have to be made out for any manufacture or production, these documents being in the form of work cards or sheets, upon which all the information an estimator desires is to be found. Within a well-organised company the information upon these documents is well tabulated, stating in detail not only the machine that work is being done upon, but also the type of work, the quantity, type of material, operator's name, etc., all detailed, not only so that the worker can be paid, but also that the company can find out what work has been done in any given time and its relevant cost. Having established that the information required is in existence, where does it go? First and foremost the usual passage is to the wages department, then on to time study, where in many cases the information is collated against a particular job for comparison with the estimated times. Eventually a report is made upon the comparison, which either gets pigeonholed or circulated to no avail, the last people to see it being the estimating department, if at all. What information should an estimator have and in what form? Every industry will have a different approach and a different requirement. Where machine operations are concerned the estimator will want to know the unit times for specific types of operations and the work form. The real questions are, how am I going to collate the information in order to make it readily available and what type of operation is it? Under what heading should I tabulate it — under the diameter of the hole, the thickness of material, or the type of material — what information do I actually require? Gathering information and data is not a simple matter. One can collate all the data, etc., available, but unless it is correctly compiled into a useful form for a particular estimator's needs, it may as well go into the nearest W.P.B. Besides compiling the data and information into a useful form, it is also possible to derive a chart or graph from the results. The chart shown in Appendix 1 was derived from this type of information. All data and information is generally referred to as historical data and if accurate can form the basis of any accurate forward estimating. There are, however, drawbacks to this type of information, such as human failings; basic documents therefore have to be carefully analysed. Which operator was it, what is the operator's experience, has any false information been booked? All these can cover shortcomings and in some cases greater remuneration has been obtained for the operator by false or inaccurate figures, either intentionally or unintentionally. In this field the experience of the estimator will tell; has the operator or the shop floor pulled a fast one over time study or wages? One does not take all information and data at face value, not until it has carefully been evaluated and when all the relevant facts have been extracted for use. One maxim for the estimator is, Don't believe everything you are told or see, it could be rigged for your benefit. One thing to do with all the information and data once it has been collated into a useful statistical form is to make it form the basis for an estimator's manual for the company's work or products. In the past it has been simply the fashion to gather and collect information like a magpie, relying upon one's memory as to where or what it is, although at times a manufacturer has been interested in collecting data for use in data sheets; otherwise individual estimators have merely pigeon-holed it. This, of course, is an extremely bad habit, although it often has been used by the estimator as a means of becoming indispensable, which of course is a complete fallacy. Any estimator who still continues to do so is doing himself and his colleagues a disservice. It is, however, unfortunate that the need for this type of safeguard, as a protection for oneself, ever arose. Each of us has in the past been guilty of hogging information to oneself for this specific purpose, and this has no doubt helped to keep estimators on a low plane, as management have not been able to appreciate the full depth of the estimator's work. Management only see the tip of the iceberg, not the mass of work that goes on beneath. It is only the final figures of the estimator's work that they see, not the mass of calculations that underlie them. Subsequently management have not been able to evaluate the job as they would have done had they known all the relevant details. True forms of estimating have not been accomplished in the past. Had there been intelligent discourses on the various ideas and methods that we know of today, there would have been, at present, a band of highly trained estimating personnel pursuing a well-defined estimating policy and functional system. Being trained, well-qualified and informed, the estimator's position would have been safe. It would have been accepted and respected as one of the functions of management.

We have, however, collated and analysed for our own particular needs a great quantity of data and information. This we have decided to make into a form of manual; but what shape should this take? Any estimating manual must be easily readable and should not contain any technical jargon purely for its own sake. It should also not contain any specialised information which may be necessary for an estimator's work but which is readily available, for example in a manufacturer's manual on a continuous welding process. The first section of the manual should be a check list which is applicable to the form of work which is being undertaken by the company concerned. An engineering company manufacturing component parts will have an entirely different list to that of a company who is engaged upon capital projects, or a steel foundry; each will have a list to suit themselves. A short example of the sections that may be contained within an estimating manual could be as follows:

1.0    MATERIAL
2.0    MANUFACTURE
3.0    FINISH
4.0    FINAL COSTS

Each section will contain subsequent check items:

1.0 MATERIAL
1.1 Type: Is it brass, steel (mild or stainless), copper?
1.2 Specification: Normal, special or any other peculiarities.
1.3 Availability: Do we stock it? A list of all normal stocks held could be advantageous (any materials which are normally used but which have extended deliveries).
1.4 Tolerances: Limitations. Are the tolerances on the material too tight or can our machines cope with the thickness, length, etc?
2.0 MANUFACTURE
2.1 Under this we can list any methods, limitations of machines, labour or vital points regarding manufacture which are peculiar to our own company's manufacturing methods.
3.0 FINISH
This should be a list of all the normal forms of finishes undertaken by the company, its associated companies, or any outside finishing companies with whom we deal, stating any limitations regarding the size of the components or the types of finish undertaken by the various companies.
4.0 FINAL COSTS
When manufacturing any article, there may or may not be service charges of one type or another. These can be either drawing or design charges; also there are other possible charges such as royalties; all of these additional charges may or may not be charge-

able to the article. Where there are such charges, the method of recouping them should be clearly stated.

Reference to the check list by the estimator will enable him to become immediately aware of anything that may have been left off of the estimate. Any check list within a manual should be given considerable thought as to its construction. If management should suddenly find themselves without the services of an estimator, it should be a simple matter to pick up the essential points by using the check list. This could happen, particularly if a company is small and has only the services of a single estimator. He or she could be suddenly taken ill, or even involved in a fatal road accident. It is only at times like these that management becomes aware of the value of the estimator. With an estimating manual, however, it is possible for the company to carry on, even if imperfect or slightly inaccurate. Having a check list within an estimating manual makes it possible for the estimator to refer to it in order to satisfy himself and establish that he has covered all the salient points. Many estimators will say that it is unnecessary, we are competent. But we are not. Should we be under the weather with even a slight cold, we can easily forget. We are all human and a simple estimate could turn out to be too simple, so we leave something off. It is a simple matter to omit something. It may be of little consequence, but it could be something important, and thus be a means of making a financial loss. It would be safer to get into the habit of referring to a check list rather than to rely solely upon one's memory. Following the check list there should be in the manual specific pages of information and data relating to the tangible manufacturing processes within a company, all to as great a detail as is considered to be necessary. For example, all the power presses within a company could be listed with their location, limitations, performance, power and any other relevant data. Subsequently a manual could be built up slowly stage by stage into a fully comprehensive document, remembering always not to add just for the sake of adding. All data and information within a manual must be related to the estimating function, that is of use to the estimator. Irrelevant data is a complete waste of time, as is also the collection of data which becomes obsolete in a very short space of time.

This frequently happens when one incorporates piecework prices into the manual. Times yes, but prices no, unless these can be of advantage for a long term. Updating money by the simple expediency of multiplying by a percentage, in the longterm is of little value, as it is usual to multiply by a slightly higher or by the nearest percentage each time, resulting in a gradual build up which becomes extensive and accumulative. Over periods of time, unless there is a change in a machine or improvements made to it, the operating times change very little, unless there has been a change in the quality of the operator (a reason for studying human relations and the output of individual operators). An estimator's manual, therefore, can become an extremely valuable document for either the individual estimator or the department. The reasoning behind having and maintaining an estimator's manual in the first place is to be able to obtain some form of coherent conformity for the estimating function. It is possible with related data and information being established that all estimates will follow a distinctive pattern which is correct for a particular company. It also follows, providing that all the information, etc., is factual, that the estimates will be as accurate as it is possible to obtain. Also the time taken in doing the estimate would be cut down as any basic element would be obtained from the manual.

Another point to consider, particularly as many companies are now in the Common Market, is that a list of foreign technical terms and their translation should be built up, as this will, in turn, cut down the number of times that drawings have to be sent away to be translated. With a list of the more common terms it would be possible for an estimator to be able to understand most drawings, as terms on a drawing are usually short and to the point. For example, the German "Ansicht in Richtung A" simply means "View in direction A", the Swedish "Putsa efta svesning" means "Clean after welding". From these two examples it will be seen that if a list of translations is drawn up, any company can cut down the length of time it takes to translate a drawing and also save the cost of translation. In Appendix 2 there are additional examples of these translations.

An important section of the estimator's manual is the day-to-day figures required by an estimator for the formulation of manufacturing costs, particularly those relating to labour and overheads. These within the engineering industry relate to individual overhead rates for each machine and the related labour rates applicable to the labour working on that machine. Normal practice requires that these are updated by the accountants every six months unless there has been either a wage award or an increase in values, which in turn could alter the overhead rates. The determination of when these should be updated should be agreed upon between the accountants and the estimating department. Because of the administrative necessity of updating these costs, this is done between the half-yearly accounting periods, which again allows the accountant to have a good idea of current values. By keeping all this information within an estimating manual, one is able to bring all the relevant data and information together, which is critical for the estimating function. An estimating manual should be tailor-made to suit a particular industry or company. By putting forward these ideas one hopes to stimulate thought within an estimator so that he can become more efficient.

Appendix 1 shows a chart which has been made up from data and information relating to cutting off lengths of cold rolled section on power presses. By the evaluation of a considerable number of figures for all forms of section and by working up graphs it was found that the relevant figures were the width of strip used and its thickness in relation to the length. It should be appreciated that charts and graphs can be prepared for almost every application, providing that one has sufficient historical information and data available to back up the factual presentation of a chart. One point to remember is that any deviation from the general theme which may or may not occur can arise because there is an outside influence upon the normal manufacture of the item in question. The first stage in the formation of any graph or chart is of course the statistical analysis of all available data and information. Upon this one has to decide whether there is sufficient detail available or not and that it falls between the relevant parameters.

The greater the amount of historical data available, the more accurate will be the prognosis and it will probably be found that it is better to formulate a graph and plot the data upon the graph, finally analysing the results, than to attempt to form a bank of data. With this initial graph one can determine whether to adhere to a graph form, plotting curves, or whether to make a formal chart or nomograph. At the same time it may be decided to have a series of graphs or charts to suit different arrangements. There can be no hard and fast rule, each will be dependent solely upon itself. The individual estimator is the only person who can judge what is best. A major goal for which one has to strive is to be able to arrive at as great a degree of conformity and accuracy as possible. When one starts to plot a graph, all things being equal and there are no abnormalities, a pattern will generally emerge. From this pattern one will have to formulate an opinion as to its consequential usefulness in the estimating function. The pattern must be accurate and conform to a basic requirement of the estimating function, as at first various patterns will emerge and with analytical probing a suitable related pattern will inevitably be found. Should a definite curve form emerge, one can safely assume that any other relevant figures will fall upon the curve in their related position. Each estimator will have to judge the accuracy of his findings and decide whether they are of any use to him in his own estimating function, and that any graph should be made to fit a positive criteria that he has set for himself. This, in time, will show up the true professionalism of the estimating function. It is the estimator who will know where in the statistical analysis of the data that the areas of natural phenomena occur and where the time of the day or the change of operator has had an influence upon the data provided. This is why any analysis must be correctly and statistically carried out by well-trained personnel. The variations or deviations may or may not be valid. It is the estimator who will know what should be included in the figures he is analysing. Variances or deviations could be because of a singularly small batch run, or alternatively an excessively large one, being included in the particular data being analysed. It is important that whenever possible the data itself is for the particular operation which one wishes to analyse, and that the operation is carried out on the same machine or machines which have relative speeds and size. One cannot simply analyse a range of data over a series of nondescript power presses, all of which vary in size or power. Care must be taken in the actual collation of data and information obtained from any documents from any source; these must be correctly assessed. One has to decide how the information is going to be used and the purpose behind it. Merely collating data is a waste of time. What one does with it is the vital question. Many estimators merely collate data under separate headings, filing it away and hoping one day that it may be useful. Having gone so far why don't they ask themselves how can it help them and their colleagues. If one formulates their ideas into any chart form, they should study whether it can take on a more progressive form such as nomographs. Mr. S. Kirkham of CIBA Geigy Ltd. has given considerable thought to this form of extension and in the field of capital cost estimating for the installation of pipework has developed nomographs for fabrication and material costs. Appendixes 3 and 4 show examples of these nomographs (reproduced by kind permission of Mr. S. Kirkham). One must not, however, assume that the formulation of charts and graphs is an easy matter. This is not so; every chart will take considerable time and study to formulate. It may appear that the time spent in analysing, formulating and assessing the information is out of all proportion to the benefits that may be derived from the exercise. However, it is necessary to weigh up all the factors, the benefits to be gained for training purposes, replacements and time saved for stand-in estimators whilst the estimator is on holiday or away ill. It may be that the time in working out an estimating formula may be so slight that to collate the information and data and tabulate it into charts may not be worth while. It may be that a simple curve is the answer and it could be the reverse and that from the data and information one may well be able to work out a formula. The short-sighted view of the estimator being indispensable and that his expertise is so great that he does not need any charts or graphs, must be overcome. Any idea of making oneself indispensable as a means of self-preservation is a fallacy and has to stop. If one is efficient, management will not change efficient staff simply for the sake of change. In the field of estimating, management require a standard of work which is reliable and changing staff about does not achieve this. Any single chart or multiple of them or graphs are all strings to the estimator's bow. It is the ability to carry out an efficient estimating function that really matters. Anything that will reduce the time factor and be of help to the estimator in being able to formulate a quick accurate answer has to be good. Estimators should get into the habit of making things to help them, as the field of engineering is too large to be able to analyse all the many facets and present them as it were on a plate as the best answer to a given problem. One therefore cannot give any absolute proposals, as to what graph, chart or formulae the estimator should make or use, but each estimator will, upon his own assessment, be extremely surprised at what can be done in this sphere. Most estimators only give thought to the collation of information and data, not that further applications can make estimating easier still. There are many books upon methods of statistical analysis, and any estimator who is unaware of any of these methods can look them up in his own local library, or write to his association or institution for advice. We all have to decide what is best for ourselves and do what work is necessary. It is also of extreme importance that any chart or graph is clearly marked as to its individual application. Too many times it happens that when information or data is given it is found that one does not know to what the figures or data refer. There is also the possibility of limitations as to their usage. All this should be clearly marked and defined. The main purpose of them, in the beginning, is to save time; it is the reverse if one is unable to make use of them because one is not clear as to their proper function. The main function is to save time, become more accurate and achieve a certain degree of consistency within the estimating field. In order to make these as perfect as possible it might be found that additional information may be gleaned from outside sources. These sources of information can usually be found in any of the following:

1. Catalogues.
2. Trade journals.
3. Advertisements.
4. Trade exhibitions.
5. Trade directories (classified).

## *1. Catalogues*

The question of whether catalogues are a useful source of information and data is totally dependent upon the industry concerned. Collecting catalogues in some industries could be a complete waste of time, owing to the fact that the catalogues are merely pictorial representations, not a technical asset. It is important that a centralised catalogue system is arranged rather than individual estimators having their own drawers full of catalogues which they feel may be necessary. The drawback to any data that may be extracted from catalogues is due to the fact that many companies are too optimistic about the performance of their materials, equipment or machines. It is in these cases essential that some credence should be obtained for the claims. It must also be remembered that some are always susceptible to printing errors.

Each estimating department will be collecting the type of catalogues that they themselves require for their type of machines or work.

Many catalogues contain considerable technical information which could be used within the overall estimating manual, or part of a specialised technical indexed system. One reservation that should be made whenever sending or asking for technical catalogues is that most companies are always willing to supply them, but will almost inevitably follow up the supply with a call by the technical sales representative. Collecting for the pleasure of obtaining catalogues could have the disadvantage of not only wasting the time of the technical representative but also your own time in having to see him.

## 2. Trade Journals

Trade journals are an extremely valuable source of information, as through trade journals one is able to keep abreast of any new innovations which are taking place within industry. Unfortunately most estimating departments are either left off the circulation lists of these journals or placed at the end of the list. Any relevant information contained in them is found out at a date which is much too late for it to be useful. Estimating departments should be placed near the head of the circulation list or have a separate copy, so that they can make use of any innovation at the earliest possible time.

## 3. Advertisements

Advertisements of course appear everywhere and in all mediums, in particular trade journals and magazines. There are also the advertising magazines which are distributed to engineers free of charge. These advertising magazines contain a wealth of information and data which can be extremely valuable to the future work of the estimator, helping him to keep up to date with all new developments within his field, and the availability of tools, machines, materials, etc.

## 4. Trade Exhibitions

From the point of view of the estimating department, management do not appear yet to have been convinced that these exhibitions can be of use to the department. It is, however, essential for the exhibition to be directly connected or associated with the work of the estimator. In most cases there is, besides the fact that the estimator can obtain first-hand information re new developments, a secondary fact that estimators can make valuable contacts which would be of assistance in the future. The fuller extension of the exhibition where there are one-day or more conventions or seminars should also be examined as a source of training for the estimating department personnel.

## 5. Trade Directories (classified)

Although these are a must for all organisations, they are usually found in the purchasing department where they stick, any reference to them having to be made in that department, the purchasing department jumping to the conclusion that the estimating department is trying to do the purchasing function, when these directories are being perused by the estimating department personnel, whereas all they are doing is examining new fields or technical possibilities. Either the estimating department should preferably have copies of their own or access to the latest copies with the old copies passed on to them.

From these few remarks it will be seen that considerable extra information and data can be collected from external sources, providing one keeps a tight rein upon the responsible person, ensuring that they are not becoming magpies collecting for collecting's sake. This is where an intelligent estimator realises the parameters of his particular industry, where to start and where to stop, and in doing so makes a considerable contribution to his work. All these contributions help the estimator to keep to the forefront of his market, as most estimates are for a considerable period of time in the future. To cover for this the sales department usually adds the words "price is subject to conditions and fluctuations in the cost of labour and materials at the time of despatch". This is represented by a percentage contingency which the estimating department has to accept, whereas most customers would prefer to receive a firm price. These fluctuations in costs during the past few years have been exaggerated by the inability of the steel trade to maintain steady prices and worse still the inability to maintain any resemblance to a uniform flow or delivery dates. The desirability of being able to give a firm price places a burden upon the estimator in respect of all these variances and fluctuations in the price of materials, overheads and labour. It is essential that the estimator forms as near as possible an accurate assessment of what these are likely to be. To some it would seem to be a simple matter of adding a 10 per cent contingency. This contingency, however, in its travels becomes the estimator's cover for any errors or omissions that he has made, not a cover for future cost rises. It is, under these circumstances, a misnomer to call it a contingency. It would be better if it were called a tolerance. As a tolerance this would be questioned, a tolerance for what? It would not be taken off the estimate so readily by other departments. If the word tolerance were used, particularly if associated with a period of time, it would make it clear as to its purpose on the estimate, not that it is a cover up for errors or omissions. With the help of the purchasing officer they should also, between them, be able to make a table of rises in all normal major purchases of materials which the company makes over the years. These, if correctly tabulated and graphed, would show the variations and fluctuations over past years, which if shown against a background of trading within the market for these years would give indication of the trends that have and may take place. Against these, one can make an intelligent appraisal of the present position and what the trend is likely to be over the next period of time. Some trade journals make this easy for the estimator, as they publish yearly assessments of the possible rise in prices. An example of this is the figures given by the *Food Engineering Journal.* These are:

*1970 Prices will Rise*

| | |
|---|---|
| Ingredients | 8.0% |
| Process equipment | 7.6% |
| Packaging equipment | 7.5% |
| Plant equipment | 6.6% |
| Packaging materials | 4.8% |

*1971 Prices will Rise*

| | |
|---|---|
| Ingredients | 9.0% |
| Process equipment | 6.2% |
| Packaging equipment | 6.5% |
| Plant equipment | 8.9% |
| Packaging materials | 5.8% |

*1972 Prices will Rise*

| | |
|---|---|
| Ingredients | 12.0% |
| Process equipment | 7.1% |
| Packaging equipment | 7.5% |
| Plant equipment | 11.5% |
| Packaging materials | 7.0% |

The lesson one can learn from such figures is that the trend appears to be a rise of 1–2 per cent per annum, accumulative upon the previous year's figure of total costs, with no indication of a drop in the future, rather the opposite that the rise curve could become steeper. From this example it will be seen that very valuable information can be extracted from sources outside the company, particularly when related to the trends as they effect prices within industry. Keeping a constant check upon the escalation of prices provides the estimator with a barometer from which to base his future predictions. It should also be remembered to keep these as factual as possible, because over a period of time the value and concept of money changes; if one takes a look at the rise in wages over the past few decades:

1939 average wage per week £2.50 a rise of 25p = 10%
1949 average wage per week £10.00 a rise of 75p = 7½%
1972 average wage per week £20.00 a rise of £3.00 = 15%

Now if we take the 1939 rise of 25p for the years to 1972 the average wage would be £7.00. On the other hand, if we compounded the 10% the wage would be £59.00. It will be seen that one cannot merely accept either of the two figures to extrapolate into the future; the figures for 1949 would be £4.25 and £6.46 respectively. One must take a look at all the aspects of the situation, taking into account the rises in wages in the intervening years, plotting them on a graph to obtain the correct solution to the past problem. However, for the future one can only surmise future trends based upon past experience. How has the market effected the wage policies of the past? Will this effect wages in the future to the same extent? Interpretation of this will depend upon the experience and training that the estimator obtains, as besides referring to ratios of market effects upon labour, materials and overheads, there are also other check ratios. Within engineering estimating fields, whether for capital cost estimating or mass production of components, an estimator can always develop a general check ratio system which will give an immediate indication whether his estimate is within normal parameters. If we take capital cost estimating first, within the food processing industry, providing of course there are no special circumstances, i.e. a complete new building or a new sub-station, the general ratio in relation to the actual cost of the plant to be installed will be found to fall between the following percentages:

| Civils | 5–10% |
| Installation costs | 7½–15% |
| Mechanical services | 10–20% |
| Electrical services | 5–20% |

It will be noticed, however, that the electrical services has the widest range for a ratio. This is due to the fact that the plant to be installed may have all the electrical services already installed as part of the machine. It may also have extremely little in the way of electrical equipment. On the other hand it may need to be wired and fitted with all the necessary electrical components. Should we then take a bodyside moulding for the automobile industry we will find that the ratios are:

| Material | 25% |
| Labour | 10% |
| Overheads | 15% |
| Finishing | 50% |

If one develops from the historical data such a system of check ratios of one's own individual requirements, one is able, at a glance, to see whether the ratios of any new estimate fall within the boundaries that one can expect. Should the estimator be armed with this form of ratio he would be in a position to be able to defend himself against any opposition to his final price. This can become an added safeguard in the estimating function, and a proof that the estimated price is reasonably correct, and that any deviation is due to special circumstances, whether due to manufacturing difficulties or an uneconomical quantity.

Aids to estimating do not, however, have to be only concerned with the analysis of historical data and information. It is quite possible that at times it is necessary to bring together various functional departments personnel for a discussion upon a particular problem. It may well be found that one of the mechanistic techniques of management can be brought into use. (This could be the decision tree to encourage clear analysis or precise thinking.)

These tools of management should not be regarded as merely tools to be used at board room level, but that they can be used at all levels. There may be, for instance, the necessity of setting up a think tank in which there is the need for the estimator to take part. A rational framework should be set up as a matter of course, to which all subscribe. The informal approaches to the workbench or sales desk should be discouraged. Where a definite answer is required, as when one is cornered or smells trouble, there is a ready-made denial of ever having done anything. The value that can be placed upon any objective framework or rational examination can in the end only lead to the most optimum output being involved at the most competitive price. In setting up any aid to be of benefit to the estimating department it is good practice to involve other departments in the general scheme if at all possible, in order to obtain their fuller collaboration and awareness of the estimating function. As estimating functions become more sophisticated, some thought should be given to an overall indexing system; so that time can be saved in the retrieval process. Any aid to estimating, as with the estimating function itself, is dependent upon the means of communication that takes place. The four recognised fundamentals of communication are:

1. That communication is an expectation.
2. Communication and information are totally different; but information presupposes communication.
3. Communication is involvement.
4. Communication is perception.

In simple terms these fundamentals mean:

1. That one must be sure that the recipient expects to be able to see and hear what the communicant desires to get across, otherwise one cannot communicate.
2. Communication, therefore, is personal, whereas information must be devoid of any personal emotion. Information is specific, however, it presupposes communication. Communication must be clearly understood by the communicant and the recipient.
3. Communication is involvement. All communications make a demand whether it is to do, to believe in, or to become somebody. Therefore the recipient becomes involved.
4. Communication is perception. Unless there is someone to hear what has been stated or heard of what has happened, no communication or perception has taken place.

Communication, therefore, becomes a vital part not only as a method of contact, but as a technique of being clearly understood. It is easier to become understood when one is talking on the same plane as one's peers, but communication takes on a different light when one is talking up or down in the organisational structure. The mode of communication invariably changes and the meaning of the information takes on a different form unless it is being considered in the same context and understood to be so, a reason why we should have estimating standards and functions clarified.

Aids to estimating take on many forms, and another practical aid is the tabulation of all the various flow systems that can be made within the company's production layouts. In order to do this the estimator must become extremely intimate with all the various production layouts within a company, those which are fixed and those which are flexible. Basic flow patterns are shown in Appendix 5. An estimator should be able to relate these to particular forms of work, or combinations of the same or other flow patterns, whether they are horizontal, vertical or inclined. A knowledge of all the possible variations will enable the estimator to be able to derive any intermediate handling times for any phase between operations, whether it follows a flow pattern or even a broken chain pattern, i.e. work placed on machines which are at different points within the factory. Many of the handling times for the intermediate handling phases can be predetermined and therefore can be readily available for any estimate, for any particular flow pattern.

Another area which needs to be more clearly established, as an aid to estimating, is the question of delivery time. Too often this is left to the sales department who have a standard set of delivery times, which are only altered when it becomes essential to alleviate the heavily extended deliveries that are taking place. Through statistical analysis most companies can accurately determine the percentage of enquiries which they are likely to obtain, and through this analysis and knowing the present state of the order books it is possible to establish a reasonably accurate forecast for future deliveries, rather than relying upon a standard set of delivery dates which apparently seem to work. Delivery dates in industry today are of paramount importance and there must be a reality placed upon them. The old question of merely stating 4–8 weeks, 16–20 weeks as a standard practice must be done away with. The customer must know within reason what the delivery of his order is likely to be, so that he too can plan. Deliveries cannot be based merely upon the present state of the order books, but what they are likely to be in the future. Whether one obtains an order from an enquiry depends upon a combination of three elements — price, quality and delivery — all of which are relevant to the work of the estimator.

## Modulex

Another aid to estimating is the "Modulex" layout planning system which gives a third dimension to any planning. No matter what layout planning problem is being considered for the company and its future, the modulex system will give a factual representation. This representation will show whether it is possible to obtain the proposed layout under existing conditions or whether new buildings or alterations will be required. For the outlay of a few pounds it is possible to develop models with this system which may well save the company large sums of money.

The Modulex system:

1. Saves drawing time.
2. Demonstrates layout advantages.
3. Is easy to build up.
4. Allows adjustments to be made on the spot.

Estimating aids, therefore, cover a variety of subjects and each aid will have its place within the estimator's armoury. It will therefore pay the estimating department well to consider what aids it requires and how to apply them to the estimating function.

# CHAPTER 8

# DRAWING TECHNOLOGY

Before any manufacturer can deliver or supply a customer with his needs it is essential that there is adequate communication between them, in order to make absolutely certain the customer gets what he actually requires. One of the most common and accepted methods of communication is through the medium of the engineering technical drawing. If we stopped to think, had not drawings been invented, how would we have communicated our needs to another party? How, for example, would we have been able to describe the manufacture of a simple bucket, let alone the complications of a motor car. Drawings, therefore, should not be merely accepted just because they are there. They have been developed over the years, into the means of technological communication which we have come to accept today, and as such more thought should be given to their precise use. As a means of communication between the customer and the supplier, it follows that the estimator must be able to fully understand all the implications of any technical drawing to perfection. If the estimator should be unable to understand the drawing and the message conveyed therein, there can be no true estimate. Any failings upon the side of the estimator will reflect in the value of his work. It is, therefore, important that this side of an estimator's work is clearly understood. It should be clearly understood that where a draughtsman is intent upon making a drawing technically correct, the estimator is intent on whether, in fact, the object which is portrayed in the drawing can be made and if so at what cost. To this end, although drawings must be technically correct, it is whether the estimator has the ability to interpret them and translate his findings into a practical usable estimating form that really matters, and any failings in this field could be disastrous to the company.

Drawing technology is the ability to be able to translate and understand technical drawings, recognising any potential problems or difficulties within them, i.e. can it be made as per drawing. It also covers all the mathematical interpretations of a drawing and the development of any situation into any basic material requirement, for example what is the

81

development of a cone upon a flat sheet of mild steel, assuming that the cone has to be cut from a basic sheet and formed into a cone. An estimator should be fully conversant with all the problems inherent in his industry, so that he can immediately pinpoint any difficulty which may arise from the manufacture of a particular article, product or project. Draughtsmen are trained personnel in the field of technical drawing, which is one reason why many estimators are drawn from this profession, as it is important that management does give some consideration to this part of the estimating function. In the past there has been a considerable over-reliance upon personnel outside the estimating function being consulted who are expected to pick up any fault in a technical drawing. This has arisen because many of the estimating staff have been merely thrown into the position of an estimator without any prior training. This type of occurrence is to be deplored and fortunately it is dying out. In the modern progression into the age of technology, it is only those companies who have fully trained estimating personnel who will be able to survive. The basic fact remains that all estimators must be fully trained in the field of drawing technology and should be aware of what a technical drawing is, so that adequate communication can be made. Therefore in a technical drawing we must look for:

1. A title.
2. A form of reference, part number and/or reference number.
3. The scale.
4. The projection.
5. The responsible draughtsman.
6. A material or parts list (if applicable).
7. Specification, tolerances or finish.

1. A title. This is most important, as one wishes to know what the customer calls the component or article, otherwise we can get our lines crossed.
2. All components should have a reference or part number, as in many cases there are more than one of the type of article or component. We therefore wish to differentiate between the various types.
3. Scale. Without a scale being put upon a drawing we have only the dimensioned parts of the drawing to work to. If the drawing has no particular scale, we are then unable to give dimensions to any undimensioned parts of the drawing, unless we accept that the

drawing is to scale and that the undimensioned parts of the drawing are the same scale as any dimensioned part, if any is given.

4. The projection. There are two systems of projection accepted in industry, the first angle projection and the third angle projection. The first angle projection is the British system, which is being completely superseded by the American third angle system, the third angle system being now known as "The Universal Graphic Language", the difference between the two projections being in the placing of the various views (as can be seen in Figs. 10 and 11). In the first angle projection it will be seen that the main view of the

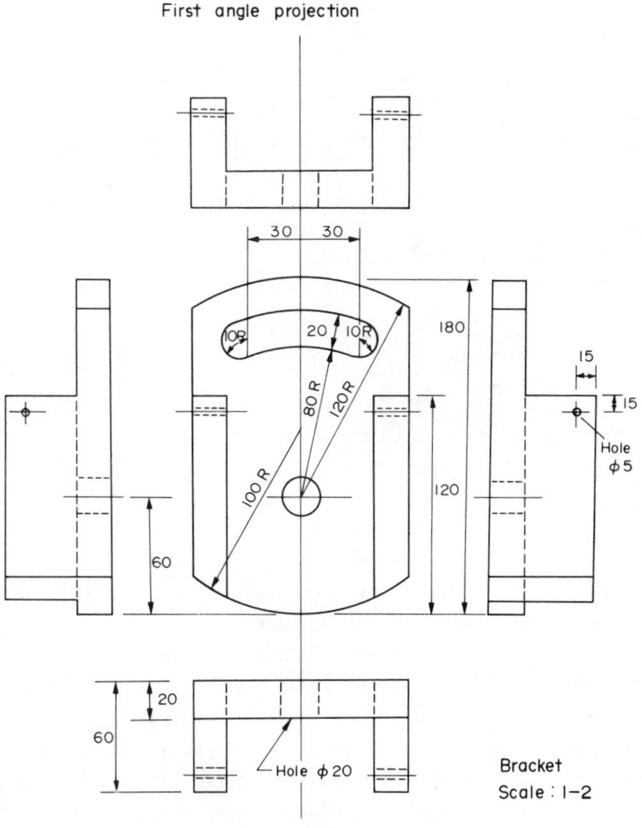

First angle projection

Bracket
Scale : 1–2

Fig. 10.

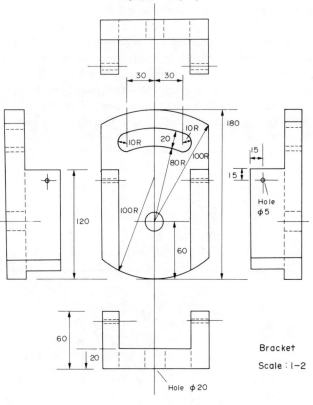

Fig. 11.

object is placed centrally, and the other views grouped around the main view. Each view, however, depicts the view of the object farthest away from the adjacent main view, whereas in the third angle projection the views shown are those of the adjacent view of the main view. Without having a clear understanding of the projections of a drawing, the estimator is unable to clearly establish, within his own mind, a three-dimensioned image of the object, neither is he able to draw an isometric view of the object to satisfy himself, should he have any doubt about being able to manufacture the same.

5. The responsible draughtsman should have signed the drawing. In doing so it is possible to be able to contact him should there be any doubt about any section or part of the drawing which needs clarification.
6. Where a component consists of more than a single item it becomes necessary at times to include a material or parts list, either marked out upon a drawing, or if more complicated issued as a separate form.
7. Any information giving the specification, tolerance or finish of the article or component is totally relevant, and any of these should be clearly stated so that there can be no doubt about what is actually required. There must be no ambiguity in any of these items; they must all be clear and concise so that they are easily readable and readily understood.

The art or science of drawing technology is one with which the estimator should be fully conversant, understanding all methods of technical drawing and the requirements of British Standard Specification 308.1964. A test of any estimator's ability to be able to read drawings would be the ability to clearly understand the drawings of the inside and outside cylinders of a steam locomotive (these are reproduced in Appendixes 6 and 7), particuarly if the estimator could work out the weight of metal involved in their casting.

# CHAPTER 9

## THE IMPORTANCE OF HUMAN RELATIONS

One of the fundamental concepts which an estimator must be able to judge, and about which he must be able to have feelings and understanding, is the field of human relations. Much has been written about human relations, particularly by sociologists, and the estimator in his own way is a sociologist. A considerable proportion of his time is taken up by working out in his mind the machinations and problems of labour. It is not merely sufficient to state that a man or woman can operate this or that machine at a particular speed or not, the empirical question is how does this or that man or woman operate the machine and what motivates them. The fact that a machine can be theoretically worked at a particular speed is not of sufficient importance. In this, the estimator not only must have studied the mechanical workings, but also the significant and more important human factor. The environment of the worker is another salient factor in any understanding of a working problem. An estimator must consider this point because this will help to determine the speed at which the worker will operate a machine. Many of the underlying grievances in industrial disputes today lie in the failure to recognise interactions that take place upon the shop floor. Some of these interactions commence because the time study engineer tries to match his figures with those of an estimate, in order to make the article at the stated price for that single section of the estimate, without taking an overall view. It must be understood that in nearly all cases the estimator is acting and working purely upon the hypothetical and not the actual of time study. This, of course, is the subtle difference, a fact which many on the outside do not clearly appreciate and why the overall estimate must be balanced and not an individual element. The estimating function must be recognised for what it is, not that it is merely a guessing game. Although working upon the hypothetical, a good estimator will be able to make an accurate overall estimate of the labour involved in making an article. This is where his basic knowledge of the workers, their environment and motivations will tell. If an estimator is unable to understand the human relations and attitudes

86

that exist between man and machine, worker and worker, worker and management, he will be unable to assess any estimate with any reasonable accuracy. Not only must he be conversant with what actually motivates the worker, he must also be familiar with the Factory Acts that govern the particular industry to which his work applies. In fact most estimators should have had considerable first-hand experience of the work on the shop floor, thereby being able to understand most of the motivations and idiosyncracies of the workers, plus the formation of the formal and informal groups. Because of this he is usually in a position where he is on extremely friendly terms with many of the workers. Workers will, in general, give assistance to the estimator, depending upon the amount of respect that they have for him. They are also normally very sympathetic to the estimator's plight and appreciate his difficulties. Many are also aware of the fact that it depends upon the estimate as to whether their future work is obtained or not. Some estimators have continuing and lasting friendships with workers. This is a relationship against the general attitude which normally exists between office staff and workers. It is also advantageous to the estimator to keep these relationships on a free and amicable basis, as by doing so he has open to him a considerable source of intimate information about the factory which is likely to be of assistance to him in his work. By the further use of tact and a little psychology he is able to extract vital information regarding the capabilities and performance not only of machines, but guards, jigs and personnel. This is not only beneficial to him as an estimator, but also to the worker and the company, as he is in a position to bring anomalies and ideas to the notice of management, particularly in companies where there are bad lines of communication between worker and management.

In estimating there is a tendency by estimators to base their estimates upon so-called actual performance figures, particularly machine time figures. These all present a very poor base upon which to establish an estimate. First and foremost it is inconceivable for anyone to expect a worker to continue to work for the whole of an eight hour shift or even for a single hour, fully extended; as a start it is more realistic to accept a figure of 50 minutes per hour for every working hour, at an average machine time, less handling times. Many companies still adhere to the old 133% performance figure, although many are more enlightened and are following the more modern concept of standard minutes. Much thought must be given to each individual problem regardless of which system is being used. All rating times are theoretically geared to enable a good worker to obtain a higher rate of remuneration for a higher output. Rating figures, however, when applied within an estimate do in many cases complicate the issue, because personnel in the various departments who have dealings with an estimate will put a different interpretation to the figures, sometimes altering them to suit their own particular needs. It is far better for any estimate to be based upon factual data, as this does ensure that a more realistic figure is used. In any event, ratings and the eventual meaning of a rating in its relation to work is always discovered by the workers, who will find some way in which to turn it to their highest monetary advantage. On the other hand, the worker has respect for the estimator who is endeavouring to bring work into the company, and it will be found that if reasonable figures have been used there is sufficient money set within the estimated costs to pay the worker a reasonable and fair wage for his labour. These figures can only be acquired by the full understanding of human problems. It has been known for estimators to quietly attempt to time workers themselves with a stop watch in their pocket. This is an extremely bad habit, because if it is found out it turns good relations into bad ones. It is also an unnecessary practice, particularly when all the necessary information of this type can be found out more accurately by the data which can be extracted from wage or work cards, depending upon the form of recording undertaken by the company. This type of information would be more factual than from any other source, particularly if it was interpolated with the data collated by the work study department. If this was done and compared with the original estimates, an accurate appreciation could be made of the capabilities of individual workers. Before any of this type of work can be carried out, there must be a steady build-up of historical data and the estimator must also be able to put himself in the position of the worker. He must be able to envisage all that would or could happen, therefore making it possible to judge and work out a factual picture relating to the problem on hand. It is essential that the estimator must have first-hand experience of working on the shop floor, particularly in relation to his own industry; not only to have worked on the shop floor but in the right locality, in order to understand the workers, i.e. can a Geordie understand a Welshman? Or a southerner a northerner?

The study of human relations between worker and management is not a modern concept. In fact as early as 1911 F. W. Taylor published in his book *Principles of Scientific Management* a list of what he regarded as principles to be followed in scientific management in order to foster a better understanding of human relations. This list is as follows:

1. That each worker should have a clearly defined task.
2. That standard conditions are needed, to ensure that the task is more easily accomplished.
3. That high payment is to be made for the successful completion of tasks. That workers should suffer loss whenever they failed to meet the standards laid down.

Taylor also listed what he regarded as duties that management themselves were to follow:

1. The development of a true science of management.
2. The scientific selection, education and development of workmen.
3. Friendly and close co-operation between management and workers.

Taylor, however, was not alone in this field, as at the same time H. L. Gantt, who had also worked with Taylor, expounded his views upon human relations. Gantt, however, is more widely known for the introduction of his "Gantt" charts. These men, however, were the forerunners, laying down the foundations of the human relationship movement, devoting considerable time and study to the science of time and motion study. Taylor, however, was laying down the basic principles of human relations within industry. These principles are clearly ones which effect the estimator today. It is an accepted principle today that each worker should have a clearly defined task. With the evolution which has taken place within industry this has become a fact; trades and skills are more clearly defined, the modern worker knows what job he is employed for, in fact he has often had his job defined by the union to which he belongs. All these definitions are of assistance to the estimator, as he is able to give a clear definition relating to any operation or task for the labour content of his estimate, these definitions being clearly recognisable by one and all. Standard conditions are also an absolute necessity, as without any form of standard conditions there would be no point in doing an estimate. Standard conditions mean clear concise operations, logical sequences and a good working environment, all of which gives the estimator a firm base for his estimate. The fact that a worker should receive a just reward for his labour is not only the concern of the estimator but also today is a recognised maxim. However, it should be clearly understood that an estimator cannot always be acting as a benevolent benefactor for the worker, at the expense of not being able to give a price at which an order can be obtained or allow the company to make a profit.

The question whether workers should suffer loss, if they fail to keep up to the standards laid down, is a question which has through time sorted itself out. All workers in some way or another suffer loss if they fail to keep to standards laid down. Each and every industry has its own set of rules or forms of payment, which result in financial losses for the worker. All this came about with the advent of the Trade Union movement, which in conjunction with management, sometimes after much industrial strife, laid down the conditions of employment within industry. One form of financial loss which comes readily to mind as an example is of course "piecework", in which the slow worker will suffer; every worker who is paid by this system is fully convinced that it is 100% orientated towards the benefit of management. Other workers are paid special bonuses for work done over and above certain quotas, all being attempts to activate some form of motivation within the worker to obtain the highest rate of output. However, management techniques are improving and much is being done towards developing the psychological approach to human relations. As the estimator is affected by the human relationships which abound within a company, he or the chief estimator should be informed of any discussions which will affect these conditions. An estimator is fully aware of the local workers' environment, as he is able to study the workers' environment at first hand through his normal connections in working and social life. He will also automatically evaluate the interactions which continually take place between the worker, his place of work, and the particular worker's motivation towards work and why he works. All this the estimator assimilates, in general quite naturally, without giving himself any particular reason why he is doing so. He only knows that when he is tackling a problem these observations can be weighed against the activations of labour. It is a gross misconception of the estimator's function not to clearly understand the implications of these relationships, and not being able to understand this is a principal reason why many still think of the estimating function as a clerical one, not as an art or science. An estimator must be able to think, judge and formulate an opinion and arrive at a justifiable answer which can eventually be proven. In this context the field of human relations, in all its varied aspects, must be considered and understood. The estimator uses his personal experience and knowledge for the estimating function and to do this he must be in complete harmony and on friendly terms with all parties within the company. This pattern of human relations can take the reverse direction for the estimator, depending upon who is his superior. Previously I have stated that sometimes the sales department and others consider that they can estimate. To counteract this the estimator has to understand human relations from these varying points of view. Knowing the importance of certain figures and their possible fluctuations and influences upon the final estimate figure, estimators at times have to anticipate interference by sales and others who are their superiors, and the estimator will either have to inflate or deflate certain sections of the estimate, knowing full well that these sections will be altered to suit their suppositions. In making the allowances, he therefore foresees the final result coming out at the figure he wants. It is a pity that some estimators have to overcome this form of irrationality in human behaviour, which only results in a bad estimating practice. But how otherwise can an estimator overcome the "bees in the bonnet" attitude of personnel who are in higher authority doing this type of alteration to the estimate? It is only by the full understanding of the behavioural pattern of the human relationship factor that an estimator can carry out his work with any reasonable accuracy. It is also rather unfortunate that, on quite a number of occasions, the estimator will find himself in the middle of what at times appears to be total open warfare between the worker and management. However, it is at times like these that the estimator will have to keep a cool head. He has to deal with all sides, therefore he cannot personally become involved in a dispute. He must be diplomatic in all his dealings, besides avoiding taking sides within an industrial dispute. The estimator should also keep clear of any internal intrigues that go on for power within an organisation; by doing so he will increase his knowledge of human relations and improve his neutral position. Through all these interactions the estimator should be able to see any area of difficulty which may result if he should estimate for a job in a particular way, and so be able to anticipate for any eventuality. A considerable number of relationships will therefore cross and recross the estimating function, but the study of human relations and the environmental behaviour of the worker is most important to the estimator. It is amazing how a little academic study can enable one to understand the worker better; also how the environment and social conditions within a company will have limiting factors which will affect the production output within a factory drastically. One of the commonest ways to affect production within a company is to have a bad canteen. The initial losses of production which can be critical are for the half hour before and the half hour after the luncheon break. It also has other effects during the working day, anything from a grumble to gastro-enteritis. A good canteen, on the other hand, will have the opposite effect and make the working staff more contented. Management should study the interactions of social conditions within their company for effects upon production.

There are also many other factors which affect production: the one-upmanship of status symbols; promotion being given to the wrong man; credit to the wrong people; all differing factors which can combine to give a considerable amount of discontent, leading to inefficiency and loss of respect for those in management. Workers and staff should all have respect and have a feeling of being wanted. To this end it is pleasing to know that some companies are pursuing progressive policies. It is critical to all that the worker has a sense of belonging, as this not only affects his efficiency but also his motivations. The right to survive is the generally accepted version of why a worker works, and that this is his principle motivation. This fundamental theme has led the worker through to the present day, where collectively he is in a far more secure position than he has ever been before. The security he wants is to be able to bring himself and his family through life, fed, clothed and housed, with the ability to have some freedom of leisure and the means both to pay for it and enjoy it. Freedom of leisure to some can be a new car to enjoy, to others a holiday in Spain, whilst others a pint at the local, all of which motivates the worker to work. Human needs are classified by the sociologists under three main headings:

1. Material needs.
2. Social or companionship needs.
3. Ego or prestige needs.

The human being is too complex a person to classify their motivations under one or other heading; they have motivations to more or less a degree under each of them. Material needs are those of feeding, clothing, housing; all workers and their families need to be fed. In the past, before the Industrial Revolution, the majority of workers were tied to the land and they consequently had to feed off the land, either supported by the Lord of the Manor, or by cultivating their own piece of land, supplemented by catching wild animals, i.e. rabbits and fowl (also sometimes by poaching). The workers in the early stages of the Industrial Revolution were totally exploited and had to feed the best way they could with whatever money they or their children could earn. This mere existence was a great motivation, the need to survive. Today, however, the pendulum has swung and the worker is more intelligent and enlightened, a fact that some still do not fully grasp. The cost of labour is therefore much higher than it has ever been. At the time of the Industrial Revolution there was little need for the estimator. The industrialist could afford to charge his own price according to the market at the time. He could also lay off any unwanted labour at a minute's notice, or alternatively hire labour for an hour or two as it suited his purpose. If he happened to have a good year, he may have responded to a prick of conscience and act in a benevolent manner towards his workers, but at no time did he relate his profit as a reason for giving higher wages. Another act of singular importance which has affected the motivations and the wealth of the working class today is the emergence of the woman as a source of labour, particularly the married woman. This has in effect put another wage-earner into the family, which enables the working class family to obtain those things which previously have been outside their province; so much so in fact that they have in many cases outstripped the former middle class. With this change in the tastes of the worker there came the emergence of a great consumer market, in all spheres, clothing, housing, consumer and luxury goods. It is not now an uncommon occurrence for the worker to eat out, in fact this has led to the increase in steak houses and Chinese restaurants. To dine on steak, chop suey or duckling is not now the province only of the rich and well-to-do. Those days have gone and the higher standard of living is to the worker's liking. He has no intention of returning to his former state. This is, of course, a major motivation to the worker, which covers not only material needs but also social and companionship needs. No longer does he only have friends within his own locality, but also friends in other localities, even abroad, which in some cases has, in turn, become an ego or prestige need.

Motivations, therefore, are not only found under a singular heading, but as they effect different workers so they will overlap from one to another. To all this has been added the introduction of the persuasive powers of the mass media which has helped to create the upsurge of ego or prestige requirements. No longer does the worker have to take a back seat. Mass media has shown him how to obtain his own dinghy, caravan or car. All these forms of leisure are paraded before the worker, and he finds that requiring the security of owning his own home is not enough; he can obtain more, providing the family work together, so it motivates them further, to obtain more capital by working.

All these desires have, in turn, from their modest beginnings been the incentive for workers to band together into collective unions in order to be able to bargain more effectively for a fairer share of the wealth of the nation. At times, however, the bargaining between management and unions has broken down to the detriment of both sides. In addition, quite often the friction between both sides has started at the bottom, to be sprung as a *fait accompli* upon the higher echelons, those in the middle screening the friction from the top. Estimators sometimes are in a position to be able to recognise these symptoms at an early stage. If they do anticipate that there is the possibility of interactions which could effect the course of the estimate, they should bring this to the attention of the chief estimator, so that either action can be taken to avoid the issue, or allowances made for its future effect upon the estimate. It is evident today that there must be a great improvement in human relations within industry, so that estimators can meet the challenges of tomorrow fairly and squarely, and also to promote the feeling of being wanted. As to what motivates workers wholly, this is different in every part of the country and many sociologists have written books upon the subject. The rules are different in various localities and the various trades conform to a different set of motivations. For instance, there is in some areas a marked reluctance to work any form of overtime, even when it is essential. First, because the worker feels that if he hurries and does the work on hand that subsequently he will be left with none to do later during normal working hours, thus making conditions where they are likely to be laid off. Secondly, that if he works the overtime the monetary reward is valueless because of the deduction of tax. There is, therefore, no desire to work overtime, yet conversely in other areas and trades overtime is accepted as part of the job. Today there is a marked and growing tendency for the worker to work as short a working week as possible for the greatest possible reward, in order to have ample leisure time and the wherewithal to pay for it. The intelligence of the worker tells him that he is no longer going to slave for a pittance as did his forefathers until he is 65, only to live a short time afterwards, assuming that he does manage to reach 65.

No! He tells himself that it is possible to be able to enjoy oneself before he reaches 65 without having to suffer ill health. He also knows that with the advent of adequate pension funds he will also be able to live reasonably well after retirement age. All this is only to be expected. If we are to have the advancements in technology, we must also have the advancements in conditions of work and employment. A few years back in 1938, office hours were 54 hours per week; now they are 37 hours per week and will eventually get less, coupled with an adequate remuneration to be able to live and enjoy the extra leisure time. It is also evident that if we are to have full employment we must resort to some form of work sharing. The thin line where human relations end and industrial relations begin is indefinable. There is the simple answer that human relations are part and parcel of industrial relations. This may be so where the spectrum is defined as being purely industrial in its aspect. The human relations of interest to the estimator, however, would appear to be more the narrow personal study of the worker than the wider spectrum.

Human relations will be seen to have many faces and the emphasis of human relations that the estimator has to be concerned with are those that directly affect, in any form, the processes of production. It is this particular emphasis upon human relations that the estimator has to rely upon, which unfortunately at times leaves the estimator in a position where the personnel manager has dealt with a problem which is of direct or indirect concern to the estimating function. Management executives on numerous occasions have had to act upon policy matters which directly affect the future of the company based upon figures supplied by the estimating department, which is at times kept in the dark about matters which could effect the validity of the figures. If one looks at the labour content or the overhead content of an estimate, one will realise that this part of the estimate may have to remain operative for periods of time well over six months. It is essential, therefore, that the estimating department is aware of any industrial activity which may affect the basis of these figures in the future. Whilst an estimator is responsible for this long-range forecasting, he is usually not informed that any award or bonus is going to be made until well after the event. It is therefore quite normal for the estimator to anticipate this form of happening, which is usually merely covered by a contingency, not a defined "tolerance". It is because of his special relationship with the workers which enables the estimator to have his own inside source of information. This helps him to anticipate the outcome of an industrial dispute and by doing so set his figures to cover for these eventualities within an estimate; any orders obtained are then covered by the estimate for the new labour content.

A trained estimator becomes experienced in this field and because of his position can become extremely successful in evaluating the climate of the workers towards any proposed wage structure or increase, due to his understanding of human relations as they exist between worker and management.

Conditions generally accepted as being of principle importance in wage negotiations are:

1. The capacity of industry to pay.
2. The workers standards of living.
3. The relation between the wages of the worker concerned and those of other workers, including workers in other occupations and industries.

In all these aspects the estimator is primarily involved, because if he is unable to evaluate labour rates correctly, he is unable to produce figures upon which a company can act and obtain orders in the face of competition. Human relations and the motivations of the worker loom largely in the estimator's work and even the best of estimators can come unstuck if the local football team "wins the cup". Estimators, therefore, should be personnel who can understand human behaviour, and in this respect the best are those trained from the shop floor; they know what can be done and how it can be done; it is very seldom that they can be misled on a particular piece of work.

Call it sociology, psychology or what you will, the whole question is still the personal relationship and feelings that occur between the worker and chargehand, foreman and manager, etc., the physical surroundings. These all interact upon each other. It is the estimator, on the side as it were, that is generally the only person able to see, judge, formulate and assess an opinion of what work will be done under these conditions.

# CHAPTER 10

# ESTIMATING AND THE COMPUTER

Within industry the approach given to computers is either one of the computer being the sacred cow, or where one is frightened off by the language used by computer experts, or their approach to the subject. The computer has been sold in great numbers to industry and the main function of the computer "is to do the wages". This, of course, in the past has sold computers, but is it really the computer's job? There is a great need for the computer to be harnessed to the many jobs that are still waiting to be done within industry on the technological side, one of these being the field of estimating. Many estimators have come to the premature conclusion that computers cannot be used for estimating purposes. In some respects there is an element of truth in this, as the use of computers in relation to estimating has its limitations at the present time. However, if we always look on the black side and do not look towards the future, we stand still. If our forefathers had not looked ahead we would have still been using the horse and cart. The question of the use of computers in the field of estimating has been of extreme interest to me and I feel that there is a great deal that computers can do in helping and assisting the estimator. Unfortunately as with all things there are drawbacks. Using computers in estimating is of course totally dependent upon the experience and efficiency of the estimating department, its wealth of historical data, and the manner in which it is recorded. The now age-old saying that a computer is only as good as the information that is fed into it is applicable, and it would take a considerable amount of time to set up the data banks for this type of work. In certain types of repetitive industries this is totally worthwhile and it is being done with package deals such as "PLUTO" (about which computer companies are only too willing to give full details). This type of package deal suits repetitive forms of manufacture where there are similar operations and sizes, when information can be fed into a computer about a current enquiry and the relevant information is taken out of the computer store and fed into the computer, the answer being worked out in seconds or a fraction of a second, thus saving a considerable amount of time for the estimator. Some considerable thought and time has already been given to the application of computers in estimating, and besides the use of the computer on repetitive work, it has also been put to good use on pipework and motorways. (The Imperial Chemical Industries has a complete ISOPEDAC system for monitoring pipework costs and the system is designed to cover all aspects of pipework detailed design, by the use of the computer.) In capital cost estimating there is the daily use of the CPA, PERT planning and PERT cost systems on the computer.

A major drawback for engineers is the fact that computers are invariably under the control of the accounts department and it is with total reluctance that engineers obtain any computer time (if ever at all). Companies may well be advised to ascertain whether they could let engineers have a satellite to the computer. When one examines the capital cost of running an engineering department, which is normally only a fraction of what is spent on packaging, advertising and marketing, it does appear to be a pinchpenny attitude that is given to the engineering department in the budget. The cost of having a computer satellite within the engineering department would in the long run pay for itself, but it would have to be a long-range return, as a considerable amount of time would elapse in the setting up of the data storage banks and carrying out the great number of trial runs on costs and estimates that would be necessary before obtaining any satisfactory results.

Computers were made for technologists, and it is within the field of technology and engineering that they should be used. As previously stated, it is rather unfortunate that in order to get computers off the ground, as it were, and into production, that they had to be used for other purposes in order to make them saleable, the principal argument being to save manpower, when it is highly debatable whether any company has made any financial savings purely upon the basis of manpower saving when using the computer for wages. A modern trend is against individual companies having their own computer unless they are large enough to be able to afford them. Instead, the agency is coming forward with computers of high power, to which their customers have access, the advantage being that these companies have access to a computer of much larger power than they otherwise would have, without having a heavy outlay of capital. Computers are being applied more and more in the engineering field, for example on mathematical problems associated with the curvatures of optical lenses. These problems are being worked out at speeds hitherto unknown and the result is that more and more complex lenses can be manufactured from the basis of the computer's answers in a far shorter space of time. It must not, however, be accepted that the computer will immediately answer all of our problems. On the contrary, it will present problems, problems which still have to be resolved in any industrial field.

As with all things, there must be a start and an essential basis for any estimating is the storage of historical data. Many companies may well be advised to think about installing a quick retrieval system of information for estimating purposes. This would be the data storage bank of a computer system. The amount of information which could be stored and retrieved quickly could be an important step towards putting the estimating department on the trail of a programme for estimating on the computer. One should not imagine that putting the estimating function upon a computer will be the answer to the problems of estimating. There will be many teething troubles, but starting with an efficient retrieval system will allow one to progress stage by stage into a full programme. The mere use of the computer as a retrieval system would save time for the estimating department, in being able to obtain information and data with a far greater speed and accuracy than by normal manual means. This is to be desired, as any system that will save time in the preparation of an estimate should be followed up and if suitable implemented; time is extremely valuable in the initial stage of preparation for any estimate. One of the many arguments against putting estimating data upon the computer is the fact that prices quickly get out of date and as such are worthless. This is, of course, true. It is, however, a point about which I have had much thought. The answer is not at all simple, as it will depend upon what a company manufactures, whether it would be possible to use a system of units or not: to instigate and develop a system of units, units being a standard for the same element, which could be translated into a value, either by conversion or by a mathematical formula. At the present we call everything by its correct name, even when there is not any reason to do so. If we are going to use modern techniques, we must be prepared to think and to use these techniques in unusual ways if necessary, even if we have to invent new names for elements, etc. Within the estimating function we must be prepared to break down functions into areas which can be common to any individual element, but which still remains recognisable. Should we arrive at a system where we have recognisable units which can be interpolated with other units; no matter how variable, we shall have arrived at a system of estimating available for the computer for our own particular requirements.

Once having discovered a system, one should not immediately jump to the conclusion that it can easily be translated into a computer programme. One must at first obtain specialised knowledge whether it is viable or not and able to be translated into a computer programme. Limitations for computer programmes, of course, are set by the equipment on hand and although it may be found that the form of estimating itself cannot be programmed for the computer, there is still no reason why the computer should not be used. As statisticians know, the computer can be used on equations for multiple linear regression (to obtain the basis for graphs) which can be used for estimating purposes. A study of statistical analysis will again show which forms would be best for the individual estimator. It should be realised that the use of the computer in the estimating field is still in its infancy, but that there is no reason why it should remain so.

A complete programme for estimating on the computer is a thing for the future, as there is still a considerable amount of research to be done in this field..We have therefore to continue to probe into this area and hope that one day we shall be able to come to a satisfactory conclusion.

# CHAPTER 11

# CAPITAL COST ESTIMATING

The field of capital cost estimating is extremely diverse and in general covers the cost estimating of capital projects, which are generally classified either as a replacement, product or process improvement, or diversification into new products or processes/expansion of existing lines.

## Replacement

Capital equipment has to be replaced from time to time as it wears out. Sometimes a breakdown occurs and a replacement project has to be rushed forward with the greatest of urgency, according to the particular failure or breakdown, giving a minimum of time for the estimator to estimate costs or the project engineer/works engineer time to plan the best method of replacement.

Where the failure has been unforeseen or neglected, the cost of replacement can be high in relation to normal replacement costs. Failure of a 4000 h.p. d.c. electric motor, for example at a rolling mill, will have a higher rate of urgency owing to the loss of production, than a single power press amongst many. Under normal conditions replacement projects are planned for, replacing production equipment which is nearing the end of its life, when the production of the product has to be continued; normal policy is to have these replaced during holidays or shut-down periods. In replacing this equipment one should take advantage of any new innovations giving either a higher production output or improvement in the product.

## Product or Process Improvement

Product or process improvement when put forward as a project has to increase production or improve the product or process with the objective

101

of increasing the company's revenue or to realise cost savings. It is normal practice when any form of capital cost estimate is being prepared to make out at the same time a cost case to show whether a satisfactory return could be made on the capital investment involved. An example of this was the savings made by an automobile manufacturer when the gear box line of the factory was automated; automation of the production line for gear boxes resulted in a minimum of labour with a higher production ratio of gear boxes. This realised an excellent return on capital, as although the gear box was made cheaper, a more uniform product was also achieved. However, in this field research, design and development costs could far outweigh any advantage which could be made; this is where the services of a fully trained estimator would show his true worth, as any quick evaluation of the costs involved would show whether there could be any appreciable savings to be made or not by implementing the project.

### Diversification into New Products/Processes and Expansion of Existing Lines

Many companies at some time or other come face to face with the need for expansion either of existing lines or the need to diversify into new lines. When either of these situations occur, the projects put forward usually involve extremely heavy capital costs; not only for plant and buildings or alterations to buildings, but also a costly programme of research and development. Any company, because of the heavy capital expenditure involved, has to take a long-term look at the evaluation of any project, in order to ensure that an adequate return on capital investment is made. Expansion in some cases invariably means the acquisition of other companies or a merger between different companies, either in order to obtain a particular market advantage or a rationalisation of resources. New products also carry very high marketing and advertising costs when launching a new product on the market. Some products do not warrant competition, because at times an intelligent advertising campaign can nullify the advantages of the rival if any, by pointing out, for instance, why change? Our quality is better, or our price is still the cheapest for an equivalent product. Too often manufacturers have been panicked into jumping onto the band wagon by an over-zealous sales and marketing force to find that the sales advantages are not there. The assumption of change for a marketing advantage is not always right.

These are some of the reasons why estimates for capital projects have to be prepared and for which it is usual to have several types of estimates worked out, from the global to the fixed, or variances between, at each stage the board of directors evaluating whether to continue the project or not. There are also a number of terms relating to the evaluations of capital expenditure, which are covered at the end of this chapter. Estimating for capital projects can cover an enormously wide field and a large number of industries, but the basic fact remains that the resultant outcome of any estimate for capital expenditure is to enable the board of directors to make a decision based upon as factual an assessment as possible. These assessments prepared for the board should cover:

1. *How* much money will be required.
2. *When* will the money be required.
3. *What* is going to be the return on the capital involved.

The estimate will give the answer to the first part. The second part is either prepared from the estimate or evaluated by the project engineer. The third part is evaluated by the commercial cost accountant. (There is, however, a growing modern viewpoint that the third part should be evaluated by a qualified cost engineer, as opposed to an accountant.)

In order to estimate for any capital project, an estimator must be in a position where he can state what type of estimate he is preparing, and also the margin of error he anticipates. This will depend upon a number of factors:

1. The time at the disposal of the estimator in which to prepare his estimate.
2. The information and data given to the estimator.
3. Whether he is able to obtain factual data for the project or whether, for example, through secrecy, he must evaluate it himself.
4. The relevance of any civil construction within the estimate. This may or may not be his concern; it could be the brief of the architectural department or consultants.

The first item is one which we are always coming up against, the question of time. This is, of course, a crucial point for any estimator: whether he has time to do a good job or not. Too often the estimator and cost accountant have insufficient time to prepare any form of logical data upon which to prepare an estimate or cost evaluation. Schemes or projects in these cases are too long in the drawing or preparation stages without the estimator or the cost accountant being involved in the work. If the estimator is brought in at the preliminary and intermediate stages of progress he would be in a good position to be able to evaluate the project more accurately when the final scheme is available for the preparation of the estimate. Under this system, the estimator would know what information and data is readily available, he would also be in a position of knowing what data he has to obtain and whether any pertinent questions have to be asked, also whether there is any specialist content or not. From this it will be seen that there could be a considerable saving of time, if the estimator is brought into the discussion stages much earlier. Whether it would be a waste of time to do so for every project or not is a question that could be asked at the time. A principal consideration is that an estimator would have the advantage of being prewarned. This would save him from being subjugated to a last minute rush each time and having to prepare a rush estimate. Although an estimator should be keenly aware of what is required in estimating for a capital project, it does not necessarily mean that the estimator should know the total technicalities of a process or project, but he should know what equipment would be required. The estimator's function is to be able to evaluate the erection or installation of equipment in addition to the cost of the equipment and its running costs. A project engineer should be totally aware of what equipment he requires and in conjunction with the estimator be able to find the most economic provision of equipment. It becomes most desirable for any capital project that a team is formed to be able to look into the project from its first concept.

Cost estimates for capital projects are of significant importance and this should be remembered by the estimator, who should not treat his estimates in a frivolous manner, as all cost estimates form the base of all figures which management evaluate and from which they make a decision, upon which rely the ultimate profits of the company, consequently the estimator's own livelihood. As an aid to standardising capital cost estimating techniques and also as an aide-memoire, the "Association of Cost Engineers" have published a comprehensive check list on capital cost estimating for the use of estimators. The main reason for having a check list is of course self-explanatory, as a check that no significant item has been left off the estimate.

The formation of capital cost estimates will vary in different industries; for example, in the chemical industry it is usual to find three types of estimates in general use:

1. The preliminary estimate, for initial project work according to the amount of information available, which has an accuracy of between 30% to 50%.
2. The definite estimate, an estimate which is put forward for authorisation or budget purposes, which has an accuracy of between 10% to 20%.
3. The detailed estimate, which is a firm and working budget, where the accuracy is between 5% and 12%.

In the food process industry, however, although similar estimates are prepared, the accuracies are slightly lower: 20% to 25%; 7½% to 12%; and up to 5% respectively. The variances can be explained by the fact that differing projects can be for various periods of time in the future, where material and labour costs can be an unknown quantity, although most good estimators do keep a record of the annual increase or decrease in the cost of labour, materials and commodities in a percentage relationship. From these it is possible to ascertain the increase in costs over the past ten years or the subsequent anticipated increase over the next ten years. The annual percentages are not a stable factor and the longer the historical period they cover, the more interesting the corresponding graph becomes, consequently the greater the skill of the estimator to interpret it. Again it should be remembered that the estimator is dealing with the hypothetical not the historical of the accountant. He is, therefore, always vulnerable.

There is, in most industries, a clear-cut division in the estimates between plant and buildings and for accountancy and capital purposes these are usually kept apart. Unfortunately in some industries the dividing line of where a building starts and what constitutes a building is a difficult question. Principally it can be safely assumed that a structure, or whatever it may be, is a building when the plant in or on it can be taken away, leaving a usable structure or building *in situ* for the use of any other plant. A bio tower, for example, is plant, because when one takes the plant out the whole structure becomes dismantled, leaving nothing for use as a building; land is also separated as a capital asset.

As a means of identification on project flow drawings, each industry or company within an industry has its own method of colour code practice, e.g. red for new machinery, blue for existing machinery, green for existing machinery being moved. In these circumstances it is not important whether this is a standard practice in the industry, as it is purely an internal matter. The important factor is that all project engineers, estimators, etc., are aware of the significance of the colour code in use. There is on the other hand the need for using the international colour code as a standardisation of symbols where drawings are sent to outside suppliers for quotations, etc., particularly upon electrical drawings, where there could be a great deal of confusion and misunderstanding. It is fortunate that since the last war much has been done with the standardisation of colour codes for services, since when practically all companies adhere to the international or British Standard Specification laid down. It would therefore be helpful where services are shown on drawings or flow sheets for them to be shown in the appropriate colour code. Unfortunately for estimating purposes, services are generally only written down as being required on the side of the drawing, without being shown where to or where from. Besides aids like these it would also be a help if the estimator could be made aware of the time of the year when the plant is going to be built or laid down, weather conditions, transport difficulties, all have an effect upon the final costs of the project. Factors like these all have a bearing upon whether work can be carried out or not, which in turn all leads to additional costs, particularly where labour has to be paid for without any work being done. Once having determined the time at one's disposal, an estimator can determine what type of estimate he is going to prepare, as at this stage he must be aware of the breadth and scope of a project. In chemical industries, estimates are prepared by using various methods of estimation, particularly preliminary estimates in which the following methods are used:

1. Lang factors.
2. The two-thirds rule (or six-tenths rule).
3. The ratio of capital investment – annual tonnes of product.

These methods normally produce estimates of between 20% to 50% accuracy. Within the food industry, however, it is usual to either estimate by comparison or by synthetic estimation to a degree of accuracy according to the information available. However, regardless of how estimates are produced they can only be as accurate as the information supplied and the skill and experience of the estimator. Estimating has been called by some an inexact science, by others an art, whilst others a guessing game, and yet it is always upon the estimator that the onus is placed. In capital project estimating some of the margin of error must be taken by either the research and development or the project engineer for inaccuracies in the information supplied or the process failing to produce what it is supposed to, all leading to rectification expenditure. However, it is possible for extremely accurate estimates to be produced and it has been known where the over-expenditure upon a £2,500,000 project was one-fifth of one per cent because of the fact that the estimator was also allowed to cost control the project. This was not arrived at by the use of a crystal ball, but by factual teamwork where everyone was in the picture and capital cost control was worked and achieved. Estimating, if correctly applied, can be almost an exact science, not the hit-and-miss affair of the past. This is, however, only as good as the estimator, the depth to which he has been trained and his own personal expertise based upon factual and practical self-knowledge. It is also unfortunate that today there are a number of estimators in industry who have little idea of their worth or capability and who are being wasted estimating for trivial work when their expertise is such that they could be more industriously employed, particularly in the field of capital estimating. To many, capital estimating is a frightening business. This I think is purely because of the amount of capital involved. An estimator will happily go on and estimate for individual articles being made for example for 17 new pence, each being made in thousands with an annual turnover of a million pounds or more, and yet he is frightened of estimating for a capital project worth ten or fifty thousand pounds and more particularly so for a project the same as his former annual turnover of a million pounds.

The skills and principles of all estimating forms are interlocked, and once mastered and understood, estimating boundaries can be crossed and recrossed. Capital estimating can become the forte of any estimator who has sufficient expertise and knowledge of a particular industry in which he has been involved, although it is a form of estimating which requires considerable "think time", where estimators must be able to envisage the completed project in their mind. Estimators for this particular field must also have a vivid practical imagination, not to be confused with day dreaming. Having previously referred to Lang factors, I would recommend to those estimators who are interested in this form of estimating that they read some of the excellent books and articles already published about this method of estimating. Lang factors were developed by Hans J. Lang, and they apply to the petrochemical industries. There are so many refinements upon the Lang factor techniques and the module factor that they could be the subject matter for a book in themselves. The basis of the method published by Lang in 1947 for those who have not heard of these factors is the following expression:

Delivered equipment cost $X$ by   3.10 for solid process plants
3.63 for solid/fluid plants
4.74 for fluid process plants

= Total estimated cost.

Since then, however, it has been generally recognised that such factors are not constant as was originally implied by Lang, but variables.

The sixth-tenths rule also applies to the petrochemical industries, and R. Williams gave this approach to capital cost estimating for petrochemical plant in 1947, proposing the exponent of 0.6 as being the method of calculation, hence the six-tenths rule. Taking the unit cost of a known capacity, the cost of a second capacity $x$ times the first, one multiplies the resultant capacity by 0.6 to obtain the cost of the second capacity.

The ratio of capital investment annual tonnes of product is rather a less accurate form of preliminary estimating in the petrochemical field. The capital cost is obtained by multiplying the plant capacity per annum in tonnes by £$x$.

The basis of most forms of capital cost estimation is the drawing of flow sheets or diagrams, which give a general outline of the process involved for a particular product manufacture; in the preliminary investigations these may or may not be merely the extension of ideas from research and development or a small pilot plant in the laboratory (Appendix 8 shows a flow diagram for storing bulk sugar). At this stage of time the annual production rate may also be hypothetical, based upon a few figures supplied by market research. As such the size of the plant can only be of a preliminary nature. Many other factors will also not be known, such as the site where the plant is going to be put down; under these conditions the estimator can only do a budget or global estimate to give management a guide as to the region in which the capital outlay is going to be so that they can evaluate the capital expenditure involved. One of the techniques employed to evaluate a project is the discount cash flow technique, a technique used to involve greater discipline in the concept of forecasting within a business.

Discount cash flow is the rate of compound interest which will make the estimated project earnings equal to the cost of the estimate, taking into account the effect of taxation and the expected change of earnings over a period of years. It is the timing of any cash flow which is of crucial importance to any company. The discount cash flow is one of the techniques of management, using the full understanding of the timing of profits which could result from an investment and the cash flow required for the investment, which gives management an accurate indication of the investments viability. Discount cash flow assesses and appraises the value of a capital project, taking into account all the variables of cash inflow and outflow. In the evaluation of capital cost projects, there are a considerable number of terms used which the estimator should be aware of, and their definitions; at the end of the book I have included a glossary of the principal terms. Some of these are well known, but I feel that the glossary will be of use to many estimators. The evaluation of a capital project is not normally worked out by an estimator but by a commercial accountant. There is, however, a growing feeling that qualified cost engineers should carry out this function in conjunction with the accountant, rather than the accountant on his own. Often the accountant is unable to grasp the technicalities involved, therefore an engineer with an appreciation of the full cost function would be an asset to all, as he could be responsible not only for the estimation of a project and its cost control, but also the evaluation of the project initially and at a predetermined future date to see if the evaluated and projected cost savings were in fact made. If not, why not? As many capital projects are extensions of the project engineer's ideas, placed upon paper without very much thought given to the actual running costs, the accountant is often handicapped as he is unaware that engineering factors have been left off which affect costs. The estimator is therefore to blame if this side of the estimate is neglected, as he should make sure that he has all the necessary information for the project. Should the estimator be unable to obtain these facts from the project engineer, this should be clearly stated as a reason for not being able to give an accurate appraisal for that part of the estimate. All facts and information must be sifted and analysed. If one is in doubt, ask the pertinent questions which should bring out all the relevant facts; most project engineers are only too pleased to discuss these fully with the estimator. However, if the project engineer thinks of the estimator as an idiot he will treat him as such. There must in these circumstances be mutual trust, respect and understanding. This only develops as both get to know each other and the involvement of the estimator in the initial stages, when the project engineer requires a great deal of help, is the time to improve this relationship. In these initial stages the estimator's collection of historical data and his technical library is invaluable and teamwork will show its true value.

Having obtained all the information and facts for a capital project, the writing out of an estimate is considered to be an easy and simple matter. Unfortunately too many give an estimate this type of approach. This stage is where mistakes are made, and there is an absolute necessity for having a check list suitable for the particular industry and a predetermined format for the type of estimate required, under suitable headings and subheadings, checking and rechecking all the time. It is also here that one goes through the whole list of the component parts of the estimate and compares each section with the flow sheet or drawings, at the same time going through every construction and working detail with the project engineer, as it is at this time when everything is being collated into a single document that individual items can come to light. Everyone may be so conversant with the general properties of the material that one overlooks, for example, the necessity that an agitator is required to keep powder free flowing. Alternatively, it could be something that has only been thought of at the last moment and the information has not yet been passed on to the estimating department. Many aspects can and do crop up, and it is at this stage that it is best for all things to be clarified, rather than later, when nothing can be done and capital has to be found or pruned from another part of the estimate. For these reasons there is a strong case for teamwork, and if this is not done at the present time it should be, or if weakly done the ties should be strengthened. Capital estimating is the evaluation of the amount of capital required to cover the cost of implementing a capital project. For this reason vast sums of capital are involved, the responsibilities are great and if the figures are inaccurate, the kickback comes twofold, one to the estimator for being inaccurate, the other to the project engineer for not putting the plant in for the authorised amount. Here then is the inducement for them to work closely together.

One aspect of capital cost estimating that will be highlighted in the future more than it has been in the past is the need for effluent plant to process any effluent or waste from the factory or plant. This is becoming more important as local authorities, river boards and government departments become more and more concerned with pollution and its effects. To those estimators who have not had, as yet, anything to do with effluent, the following are a few of the types of effluent plant with which one may become involved. Broadly speaking for the purpose of estimating, effluent is that part of the raw materials that is unused and does not form in any way part of the finished product or by-product. It is also unwanted. As such, it may be either liquid, gaseous or solid. The question arises, how to get rid of the effluent, whether it is in its simplest form, hot water, which has to be cooled, or whether it contains solids which may or may not be toxic, or whether it is in a liquid form acceptable by local authorities to be discharged through the normal sewer system. Effluent has at first to be defined, and then it has to be rendered harmless (if it is not harmless already). It may in many cases be necessary to draw upon the services of the experts in effluent disposal. Local authorities and river boards have laid down the required limits to which effluent has to be reduced, before it can be disposed of through any of their sewage works, or into rivers and canals, although there is little legislation to prevent raw effluent being discharged into the sea. Some of the principle methods of treating various types of effluent are as follows.

*Solid Effluent*

This form of effluent is either burnt or tipped if possible. If it is tipped and the effluent contains toxic matter, when it rains the rainwater will permeate the ground, spreading the area of toxic matter even to the extent of eventually reaching streams, rivers and drinking water supplies. In this case additional safeguards have to be met. Owing to the large amounts of solid effluent, some authorities are becoming extremely disturbed over the indiscriminate tipping of this form of effluent, as it is being tipped into disused quarries and pits, particularly on the Welsh border and many such tippings have turned out to be toxic. In addition, considerable amounts have been dumped into the sea, some in special containers, some without, and there have been recent cases where ships have been made to return to port with this form of effluent and the companies responsible have been made to make alternative arrangements for its disposal. All of this has led to many companies having to lay out considerable amounts of capital on costly disposal plant or on storage facilities until such time as the effluent is harmless or a satisfactory method of disposing of it has been found.

*Gaseous Effluent*

Gaseous effluents are either treated by burning, by oxidisation by using a catalyst or by washing in order to produce a liquid effluent by the use of water or chemical reagent. Where solids are present in a gaseous effluent these are removed by various methods each dependent upon the form in which the solid is present. These solids can either be removed by filtering the gas or by separating the solid by the use of centrifugal separators (cyclones), wet scrubbing or by the electrostatic action of precipitators.

*Liquid Effluent*

The majority of effluents come under this category and it is of course totally dependent upon the location of the factory and the requirements of local authorities, water boards, etc., which will determine the type of plant required for the discharge of the effluent. Also whether it is being discharged into a sewer, river or sea; what amount of suspended solids are there in the effluent and also is it toxic? All schemes will have a differing reaction from various authorities. For example, take a case where water was drawn from a stream for cooling purposes only, and before being returned to the stream, cooled to a satisfactory degree; this project was rejected by a river authority as being unsatisfactory. The water had to go through the main effluent plant before being discharged, as there may have been a possibility of leakage into the cooling system, although the system was a complete closed circuit. Yet another authority did not object to a similar system elsewhere. It must be accepted that one cannot go ahead with any method of discharging any form of effluent, without all the factors being taken into consideration and the most economical method may initially be the most expensive. As with gaseous effluent, any solid in the liquid effluent has to be removed, either by the use of screens, grit channels or settling tanks. In some cases where the solid has to be oxidated this is done by a chemical or biological method. The cheapest chemical method is by the use of bulk chlorine, using chlorine as a reagent. Biological methods are carried out by using the activated sludge plant or by the use of a percolating filter, the percolating filter being a thick bed of crushed stone or slag in which bacteria lives and works upon the effluent which is passing through it, or by the more modern method of percolating by the use of plastic packing which allows filter beds to be made to a considerable height owing to the use of this lightweight material. The fact that filter beds can be made to a greater height than hitherto means that there can be a considerable amount of saving in the area of land used, which may or may not allow savings in capital. The activated sludge system has the effluent and bacteria mixed together in a tank or lagoon which is then aerated either by the use of compressors or agitators.

When any project has to have a new effluent plant included in its cost, the engineer is faced with many variables. It will be dependent upon the type of effluent that has to be disposed of which method the engineer will choose and evaluate. In most cases it would be advisable to call in experts to assist in choosing the correct system.

Basically the engineer must:

1. Select and compare alternative systems or solutions.
2. Find out which plant is the most economical for the particular solution both in design and construction.
3. See that the operating costs are economical.
4. Find out what the maintenance costs are and over what period of time, also what is the life expectancy of the plant.
5. Find out what amount of capital will be tied up in the provision of spares.
6. In addition find out what after-sales services are available and what guarantees there are for the plant.
7. By the use of a particular system, find out whether any part of the effluent can be reclaimed.

Most of these questions will also apply to any other forms of capital plant.

In all these considerations, the engineer would be well advised to bring the estimator in to assist in the evaluation of the costs, which besides allowing the estimator to be of assistance to the engineer in the initial stages, will also make the estimator become aware of all the problems and associated problems of the whole project. This would have the secondary value of the estimator becoming totally aware of any areas where additional costs may or may not arise, owing to difficulties which can only be assessed and not realised. This is the same with the whole field of capital estimating for major projects involving considerable capital expenditure. As already stated, the more the estimator is involved in all the stages of the project, the more accurate is the estimate going to be.

Capital cost estimating is not a function that is carried out by all companies and in many of the small engineering companies it is very seldom done outside the boardroom, capital projects being determined by the directors as being necessary for expansion and the costings being done by them and the company accountant. Capital cost estimating departments are therefore generally only to be found in the larger companies and combines; where these exist some companies offer the services of the department to other companies upon a consultancy basis. Companies who offer these consultancy services for capital cost estimating also offer to carry out the overall planning and cost control. This can be immensely valuable to a smaller company who because of financial reasons are economically unable to carry a staff or department of their own.

Capital cost estimating is not a difficult problem for any trained estimator, but because of the nature of the amount of capital involved, the work has to be more exacting and the estimator more sure of himself. Estimators who carry out this form of work must be completely self-confident and also must be able to impart this confidence to their fellow workers. Engineers working with the estimator must have respect for him and also confidence in his work. There is a total need for human trust and confidence between the engineer and the estimator, otherwise the engineer will always have doubts about whether he will be able to carry out the project for the capital authorised. In all cases the estimator must be self-reliant, self-assured and above all have the ability to do his job with full technical expertise.

# CHAPTER 12

# INVESTMENT APPRAISAL

Before a company authorises any capital expenditure for its future investments, based upon proposals placed before the board, a procedure of investigation is set into motion, to investigate all the various aspects of the proposal in question. Evaluations are obtained to show what return on capital will be achieved should the proposal be implemented. The whole reason for having an investigation is to be able to obtain a number of assessments to evaluate and so select the one which will give a maximum return upon capital invested, alternatively known as the maximising of long-term profitability. It is these investigations which are referred to by management as investment appraisals and upon which the board will make its decision and act accordingly. Capital investments are naturally long-term investments, covering a number of years, upon which the profitability of the company will depend. It is essential that the board has some means of being able to anticipate what the results of any long-term capital project will be, so that the correct decisions can be made in order that the company is not involved in any form of heavy capital expenditure which could turn out to be a white elephant. At the present time investment appraisals are the function of the commercial accountant and/or cost accountants, who base their assessments upon figures given to them by the sales, marketing and production departments, with the cost of the project worked out by the engineering department estimator. There is, in this system, many areas of uncertainties where, because of the nature of the departments involved, inaccuracies occur, sometimes because of the different interpretations placed upon the various factors. For example, the statistical variances of the engineer are different from the variances understood by the accountant. There is much to be said for the modern idea of having cost engineers trained to be able to work with the accountant and by doing so be able to work out a joint investment appraisal. This collaboration should be with the full co-operation of all the other departments involved, the reasoning behind this being the theory that as an engineer he would be able to:

1. Estimate the cost of a project.
2. Evaluate the productivity of a project in relation to the anticipated sales and marketing figures.
3. Relate any alternatives or alterations of the original project to the specific area of the estimate and evaluate the effects upon the capital and subsequently the investment appraisal, bearing in mind at the same time any effects from 2.

Logic also follows on from here that as the cost engineer estimated for the project and evaluated its worth in collaboration with the accountants, that he is the person who should control the overall cost of the project in collaboration with the project engineer. With the cost engineer thus in control of the financial aspects of the project, the accountant would be informed whether capital was being spent correctly or not, i.e. obtaining equipment on one project to cover up for over-expenditure upon another. In addition, a correctly maintained cost control function would give the accountant an up-to-date picture, capital authorised, capital committed and capital paid.

In general the accepted classification for capital investment falls into one of the following categories:

1. Diversification into new products.
2. Expansion of capacity for existing lines.
3. Profit improvement projects.
4. Essential replacements.
5. Safety and statutory requirements.
6. Quality, welfare and amenity.

In whichever classification the project for investment falls, or whatever the minimum rate of return required for a particular category, accountants will still prepare appraisals for the board in which the following will be answered:

1. *How* much capital will be required.
2. *When* and in what period will the capital be required.
3. *What* is going to be the rate of return upon the capital involved.

These are the principal factors in an appraisal. The reasons are simple. The fact is that capital is required for a project and it will be required at a particular time and at a certain rate of expenditure. Usually the question is not that a project will cost £X to be installed this month, payment to be made in the following month. It is more complex. The project is spread over many months, sometimes years. Some payments are made to certain commercial practices, 5% upon receipt of order, with payments made at further intervals of time or progress, or against bills of quantities. One cannot speak of the payment of a project as a single sum of money. Assessments have to be made of the rate of progress of the project, what is to be completed at various stages, the type of work involved and when it will be logical to expect the invoices to come in. Armed with this information it is possible to inform the accountants of the anticipated expenditure for each period, which for accounting purposes is normally each financial quarter, 'X' capital in the first quarter, 'Y' in the second and so on. It is important that this section of the appraisal should be as accurate as possible. This is an additional reason why an engineer trained in accountancy methods should assist the accountant at this stage in order to ensure as great an accuracy as possible, as the raising of capital costs money. It is simply not good enough to speak of slapdash figures; a project is going to cost £80,000 to be completed in four equal payments each quarter of £20,000, when in actual fact only £5,000 will be required in the first quarter, £8,000 in the second quarter, £15,000 in the third quarter, £35,000 in the fourth quarter and the remaining £17,000 in the first quarter of the following year. Had the accountants made provision for £20,000 for each quarter, it would have resulted in £15,000 being held idle in the first quarter, £27,000 in the second quarter, £32,000 in the third quarter, £17,000 in the fourth quarter held on into the following year's first quarter. Had an accurate assessment been made there would have been appreciable saving over the slapdash figures, which would have meant additional expenditure in interest rates, besides making a difference in the appraisal itself. The evaluation of capital expenditure, how much and when required is of extreme importance and is one field that has not always really been appreciated by engineers. This form of information must be as factual as possible to keep costs of investment to a minimum. This exchange of all available facts and information is extremely vital to all parties, as it is quite normal for the project engineer to have to design plant on the flimsiest of production or marketing data, also for the appraisal to be done upon a different set of figures than those upon which the plant has been designed. Once again this is a valid argument why an estimate/cost engineer should become involved in capital appraisals, with his expertise having estimated for the project, knowing what the project was designed for, the cost of each of the individual sections and what sections are critical to individual production rates. Should different figures be implemented at any stage, he would automatically know what their import might be. It happens only too frequently for various sets of figures to be bandied about by all and sundry, an appraisal made, which bears no relation to fact, resulting in either gross over-expenditure/under-expenditure and the installation of plant which bears no relation to production requirements, being either grossly under or over production capacity. When this occurs it presents quite a different pattern in the rate of return, which very seldom ever gets analysed, as no assessment is made of the project after completion, its performance never being compared with the figures in the estimate. Should there be any criticism of the project there is always the easy way out, simply stating "Well we made some modifications, the project is therefore different", justifying the modifications on the grounds that some of the equipment was either faulty or did not come up to expectations, facts that accountants have to accept on their face value; being non-technical they are unable to query them further. If a follow-up were rigorously enforced, engineers would have to become more cost conscious, knowing that there would be a resultant check upon their work. This in turn would make sure that they would not embark upon any project in a haphazard manner and in turn would weed out the good engineer from the bad. It is the rate of return upon capital which the board is interested in, whether the company is going to make a profit or not. Somehow today the word profit seems to have taken on a dirty meaning, but as yet no company, even nationalised companies, can exist without making a profit, nor can the workers be paid. Profit is also the capital from which all future activities of the company have to be financed.

In the investment appraisal the rate of return is found out by the use of management techniques, the principle ones being:

1. The accounting rate of return.
2. The pay-back period.
3. Discount cash flow.

These techniques or methods of ascertaining the rate of return basically are as follows.

*Accounting Rate of Return*

This method measures the expected profit increase that one assumes will result from the new investment by normal accounting and budgeting techniques; the anticipated return is compared to the amount of capital that the project requires. Normal accounting and budgeting procedures take into account the depreciation of the plant before and after tax earnings. These will differ from company to company depending upon the accounting methods used by a particular company. The pitfalls or deficiency in the accounting rate of return method is that it does not take into account the timing and life differentials of a series of investment proposals and incorporate the effect of them into a rate of return.

*Pay-back Period*

The pay-back period method of appraisal makes no attempt to evaluate the return upon the amount of capital invested. It merely shows the length of time it takes to recoup the amount of capital invested, taking no account of any further profits that may be attributable to the plant after its pay-back period.

*Discount Cash Flow*

When comparing the previous methods it will be seen that they do not take into account what the capital invested in a project may or may not have earned if it had not been tied up in the project. The D.C.F. rate of return takes into account not only the present value of the capital but also the cash flow, rate of interest, tax and the time factor. In particular, the problems introduced by tax must be evaluated, as different projects will qualify for different tax allowances, all of which will require very careful study.

It is for all these various reasons why investment appraisals must be finally carried out and analysed by qualified accountants. Problems emerge which most engineers do not comprehend and which would make it a more reasonable assumption that all investment appraisals should be a more cohesive joint effort between the accountant and engineer. Failures in industry today are either being levelled at the failure of management or the interference of trade unions in management policies. There may or may not be truth in particular accusations, but many of the failures start because of failings in the techniques of the investment appraisal. Management, therefore, is not in a position to realise whether a profit in fact has or has not been made; indeed the whole of the management technique of appraisal may be highly suspect, which in turn could lead to the failure of a company without any actual realisation of the cause. Investment appraisal, therefore, is a technique which leaves much to be desired in obtaining a near 100% accuracy estimate of the return from any capital investment.

# CHAPTER 13

# RESEARCH AND DEVELOPMENT

It is within the field of research and development (R. and D.), that area of great uncharted seas, where the estimator will find his greatest challenge, an area of consequence to all management, within which lies the future of the company. The challenge to the estimator is enormous, but to one to which every job is a challenge it becomes of more interest where the challenge is the greatest, particularly so if the estimator has the respect of the rest of the team. This field, however, also brings with it the easiest method of wasting and using up capital; within R. and D. and with the method in which prototypes are built and placed upon the market, lies the crux to the question of why there are. so many losses upon the market relating to capital expenditure. In many cases the projects have never been rationally conceived or implemented. This is, of course, easy to say, but nevertheless is the truth. To be viable in the field of R. and D., team members must each command the respect of each other in the particular field in which they are employed. They should not be subjected to outside interference, which damages the quality of their work. Also all members of a team, whether only interested in a single aspect of the project or not, should be involved from the inception of a project until its completion. This will eliminate margins of error across the whole spectrum of the project. In the majority of cases the estimator will get the blame for any financial losses incurred, as his estimates will be attacked as being faulty. In some cases this will be true, but in many other cases the involvement of the estimating department is only at the last moment when it is impossible to evaluate the influence of many so-called minor details. In R. and D. the large problems invariably get answered. The unanswered problems are the minor ones. Who would have thought we couldn't have done so and so? This is why a project team should be named in the beginning and all aspects covered. It is probably totally unnecessary for the estimator to be involved all of the time on a project, but if he is in at the inception he at least has some time in which to formulate his ideas and probe into the dark areas. In the same vein, the sales and marketing personnel who are to

121

be involved in selling the result of any project should be kept in the picture, so that they will know what they are selling, its limitations, if any, and its purpose. To illustrate what can happen to a project the following true story will no doubt be of interest to some. It concerns a large engineering combine in this country which decided to go into the field of industrialised building. A company within the combine was chosen to do the R. and D. into this particular field. The company was chosen because it was already in the building industry, mass producing many items for the major building contractors, particularly in the public sector. It was therefore well placed to be involved in such a conception. It had adequate design and drawing office facilities, besides suitable production and prototype workshops.

There was also, as there still is, a considerable market for an efficient, reliable, sophisticated, integrated, industrialised building system at a reasonable price, which would wholly stand up to an adequate fire retention limit. Industrialised building systems usually work out at a price considerably higher than normal conventional methods, particularly where fire-retardent programmes have to be met, i.e. ½ hour or 1 hour fire retention. Against this, however, has to be weighed the advantage of having a building or unit erected in a much shorter space of time, when the building or unit would be earning money for a period of time long before its conventional rival, and an investment appraisal would in fact show a related cost which would make them competitive. This, however, is very rarely ever done. If it was, it would show this fact up as a sales advantage.

The original conception had to be for a robust, integrated system into which items already being manufactured, i.e. beams, joists, etc., could be employed and utilised. The system also had to be architecturally sound. Such was the brief given to the development director. There were as usual many problems, but eventually a reasonable design was produced and the estimator began his preliminary estimates, based upon all known facts. As the design progressed so did the estimates. The costs were comparable to those of any other system at the time and it was decided to go ahead and manufacture a prototype bungalow. This was made, erected and tested in the highlands of Scotland and found to be reasonably successful. It was, however, decided to make certain modifications based upon the experience gained in the design. These modifications had little effect upon the overall costs of the system, or the project, but they overcame certain weather problems. Basically it was a system of metal framed panels, which were insulated and acoustic and of considerable load-bearing properties, at the same time being extremely light and fire-retardent. Depending upon the form that the building was to take, the integral metal frame system was changed to take up any other added stresses, i.e. wind or snow loadings. It was decided to go ahead and market the system, and up to this time the capital outlay had been kept to a minimum and development costs were almost negligible. With this system, the estimator stated that as long as the design of a building was related to the system, any price would be acceptable upon the market. Points, however, which had to be considered and taken into account were that there could be extreme vacillations in the price if very small panels represented anything over 15% of the structural face, and also that if panels which were outside the normal trade sizes for the finishing surface were used the price would escalate. The design at this time was considered by the engineers to be a successful one, but there was a general feeling that in the initial stages the whole project should be kept upon a prototype basis, until some experience had been gained in production and erection methods. It had been envisaged that a few small orders already on hand should be dealt with on a prototype basis until production methods were finalised and the training of erection teams could be undertaken, before any attempt was made at mass production. Unfortunately the sales and marketing departments disagreed, saying that the market was there, and that they could obtain all the orders required for immediate mass production. They went ahead quoting prices for all forms of conventional buildings, high rise, low rise, obtaining mythical orders which would be firm if deliveries could be guaranteed and full production commenced at once. However, had sanity prevailed and only one or two of the original opportunities followed up on a prototype basis, everything would have been alright. This was not to be. "The market was wide open" and the engineer's advice swept aside. A new factory was acquired, plant obtained, production lines laid down, supervisory staff moved to a new location in another county. With communications difficult, management being at the parent factory and the production unit miles away with no firm orders for mass production, the whole conception stagnated. Continual warnings about the usage of estimating figures went unheeded, wild quotations were given to all and sundry, based upon original estimated figures for a square foot base, which bore no relationship to any quoted building. A year or more elapsed and still no large mass production orders. A few small orders were completed at a financial loss, which had they been done on the prototype basis, as at first requested, these would have been manufactured more or less at cost, at least sufficient would then have been done to have proven the design to the industry. Time would have allowed this to have been done and a correct production method could have been formulated with all costs being kept to a minimum, whereas in fact a considerable sum was spent, reputed to have exceeded a million pounds.

This was a particular example where the complete inability of the sales and marketing departments to appreciate what they were selling led to an abject failure. Had they been made to assimilate what it was they were selling; what industrialised building really meant; what field it actually covered, this would or could not have happened, but it did. The technology of the design was good and it could have been a success. This was the story of a highly sophisticated industrialised building system, that could or should have made considerable profits for the shareholders and work for the workers, but which in effect through inefficiency led to high losses and unemployment.

The field of R. and D. has in the past and at present proved to be a particular hazard to the estimator, principally because of the unknown factors that generally arise and abound within this particular sphere of operations when evolving or attempting to produce a new product. Estimators invariably are blamed for the resultant financial losses, which in many cases is completely unfair, as the estimator has usually been kept extremely ignorant about pertinent facts and he has been expected to estimate upon a broad unknown base, any objections being swept to one side.

One considerable item in the field of R. and D. is the great number of projects which are shelved. Projects are shelved for many reasons, the principle ones according to the findings of the Centre for the Study of Industrial Innovation being:

1. *Environmental Reasons.*
a. Unattractively small market.
b. Uncertainty with monopsonistic buyers.
c. Unattractive level of competition.
d. Uncertainty with suppliers.
e. Obsolescence.

2. *Organisational Reasons*
a. Lack of marketing capacity or expertise.
b. Lack of production capacity or expertise.
c. Faulty communications with associated firms.
d. R. and D. cost escalation.
e. Shortage of R. and D. resources.

These findings were based upon the analysis of work done by 4,713 R. and D. personnel within twenty different companies in the electrical, electronic, engineering and chemical industries, these companies employing 269,799 workers. The findings, therefore, can be viewed in a satisfactory light, as a sound basis of why R. and D. projects are shelved. One such project that I know of as being shelved was a knock-up kit for a Do-It-Yourself four-berth caravan, a kit which could have been put on the market on its own chassis, carrying two crates of body parts, to be towed away from the show room; all the tools that were required to construct the caravan were a screwdriver, a hammer and a spanner; these could have been included in the kit. At the time of conception this kit could have been put on the market at a highly competitive price, with a choice of internal accessories, any D.I.Y. enthusiast being able to construct it over a week-end. The reason for shelving the project was that the majority of components would have had to have been bought out, although the major part was from other companies within the group. With only 30% of the caravan being made within the company it was felt to be an insufficient proportion for the company to produce. A perfectly good idea therefore was shelved, although it could have hit the market at a vital time. It was brought out again several years later after a change in management, but it was too late; the market had been more than adequately covered. The new management would have been amenable to have split the work between the companies. In fact had this been done in the beginning it would have saved one of the smaller companies. This time it was a missed opportunity by management, at other times it is the other way around. It is, however, the sole responsibility of management to decide whether or not a project is to be shelved. Care must therefore be taken that the project not only looks good, but is good and is a sound investment. The ideas and recommendations of all departments concerned should be carefully analysed and scrutinised. It is an easy thing to be biased towards or against a project just because one's favourite department thinks the same; when other departments with other ideas and recommendations could be right, management should always try to be as impartial as possible and examine the ideas and recommendations from all angles. To be able to conceive an idea, research and develop it, then to go ahead and manufacture is a simple matter, *it only costs money*. To go further and market it and to be able to make it pay is a very different matter. R. and D. estimating has to arrive at a figure to prove whether it is feasible to go ahead, and whether it is going to be worth while to delve into a problem in order to see if it is at least possible to be able to arrive at a satisfactory conclusion. All this requires experience and understanding of all the problems that may or may not be met in any work on any project and what it will probably cost to go ahead, whether in fact it is within the company's capability. Without any form of costs, management is left in the dark hole of Calcutta, not knowing which way to turn. In order to arrive at any cost there must therefore always be a sound basic approach to any R. and D. problem. This basic approach can only be achieved by an efficient team who between them can visualise all the apparent happenings which may or may not occur, in addition being able to nominate any areas and shades of grey which are going to present problems and jointly between them being able to give a reasonable answer to the viability of any project. Only by doing so can any R. and D. project be estimated to any degree of accuracy and this only if the team has had experience within the particular sphere relating to the project, particularly the experience of the estimator. In quite a number of cases R. and D. estimates are never really true estimates. They are only basically an allocation of money and resources to the department by the board. The split up of this is decided upon by the head of the department, either by his own intuition or more formally set by an annual budget. These resources, if they should be monetary overspent, are covered by submitting additional applications, accompanied by the different reasons why the expenditure is required, to the board for the allocation of more funds. This results at times in projects being shelved for reasons of economy, when these projects would probably have had a good chance of success had they gone forward, or if they had been evaluated more objectively. It is of course a very simple matter to criticise any form of R. and D. estimating, but in reality the cost of being able to prepare any detailed estimates of any or of all R. and D. projects could be out of all proportion to the amount of work to be done. In R. and D. the simple method of proportioning funds could be quite adequate and an easy answer to a costing problem. But to visualise a large project and all its ramifications purely upon a hypothetical basis calls for a great degree of expertise, of a team operating within the particular scope of the project.

An estimator who is both competent and experienced within the scope of the projects field should be totally capable of being able to give an accurate forecast as to the eventual costs, to within limits which can only be determined by the facts and information supplied to him by the rest of the team. Each possible segment of the project should be discussed with the engineer/s responsible so that the estimator would be able to determine the form that the estimate could take for each segmented part. Some segmented parts will undoubtedly be based upon previous work done and as such will be able to be determined with reasonable accuracy. Other segments will want to be estimated based upon the judgement of the engineer and the collective opinions of the team (if possible). In unknown areas it can be said that one's guess is as good as another's. This is where the expertise of the estimator comes into being, providing that he has been trained and has practical and theoretical experience. The estimator in these cases should be able to make a reasonable assessment by being able to contemplate and visualise what work can be expected within such an area, an assessment which to some will appear to be nothing short of a miracle. However, the estimator should be able to do so and also be able to place a tolerance of limitations to the area. There is also the fact that an experienced team should be able to assess time factors which could be involved in reaching concrete decisions regarding the project. This is why any large R. and D. project should be conceived by a nominated team from the start. Even if some of its members do not actually work full-time upon the project, they should have the opportunity of being able to be in at all its relevant stages. Any form of R. and D. work will because of its very nature gobble up considerable sums of money in the pure R. and D. stages; therefore all stops must be pulled out to keep this to a minimum so that a prototype could be made at the earliest possible time to ensure that the thinking is correct and that there can be a high possibility of a modicum of success. The success of any project will of course totally depend upon the skill of the R. and D. department, coupled with the respect that it has within the company and whether it has any resources available upon which it can call. There is always at some time or another the necessity of having a model or prototype made of a section of the project to see if it will work or not. If the tool room or the prototype department think of R. and D. as nitwits, one can hardly expect to find any form of success being achieved.

Another hazard in R. and D. work is that insufficient thought or time is given to the brief or terms of reference, the sphere of the project being too ambiguous, with many subsequent changes in direction, whereas R. and D. should be absolutely clear as to their direction and purpose. For example, one does not expect the manager of a football team to come on to the field of play anytime during a match and inform a player he is playing for the wrong team, on the wrong ground. In reality this is the type of occurrence that R. and D. have to put up with. When a project has been wrongly described in any fashion and R. and D. have to restart, the costs should be written off, not as at present where they are continued and the overall costs of the project charged to another project because no one will accept responsibility for being wrong. R. and D. is of course a company's future life blood. For this there is a need for the department to be kept completely up-to-date and for the department to have personnel who are capable of change. The work performed by the R. and D. department is many-sided. It can either be dull and plodding or completely creative and exciting. A major consideration for management is the basic fact that R. and D. costs are important and that these must be planned so that the department can be treated with the respect it deserves. There must also be some understanding of the difficulties that we cannot do without. In the recent past, however, British industry has been made sightless by the ruthless cutting back of R. and D. in the massive rationalisation programmes that have taken place. Excellent R. and D. personnel have been made redundant and consequently they have had to look for "safer" occupations, being reluctant to return to R. and D. work. A company has to be correctly balanced, and to cut off all departments unconnected to immediate production is a highly short-sighted view.

# CHAPTER 14

# CORPORATE STRATEGY

As I understand corporate strategy, it is an emergence of a discretionary management technique, which involves all forms of planning, whether short, medium or long-range. It is what the company is going to do, when it is going to be done and how; it also involves the negative. To be negative in certain circumstances can be an extremely wise decision. Too many people react too quickly to a given circumstance, particularly to the action of a competitor, when by doing nothing and keeping the status quo would have been the best thing.

Today the corporate strategy function is carried out rather loosely, either from an innate intuition of the managing director down to a this year plus 5% attitude. There are many times when the annual budget is a last minute gamble of the responsible manager, answering a final request from the accountant. Where this occurs the last year's budget is copied with slight alterations and increases to arrive at a figure that they do not expect to get, but which they expect will be cut down to a figure that they want to obtain, the only realistic items in the budget being those that have been brought to their attention during the preceding few days. Very seldom is a well-thought-out budget ever presented to the accountant and I am certain that many accountants would like to see one. This attitude must be overcome. Resolute planning of budgets must evoke strategies which are positive and correct. It is the negative attitude taken by untrained management and in a few cases by trained management which is an underlying fault in this country. The attitude of continually putting forward budgets which mentally are never going to be adhered to is the effect of this system. Not only are negative attitudes to effective budgeting taken in industry, but they also permeate into the offices of local and national government. Negative attitudes are wrapped up in such a manner that they appear to be positive, and the overall planning of industry purely for political purposes and aspirations is faulty. One only has to study the Soviet Gosplan to see examples of cause and effect. Planning for narrow political ends within one country when the overall strategies are affected

129

by world-wide conditions presents chaos. Galbraith advocated the nationalisation of companies in his Reith Lectures. This is still being advocated today, although the time for doing so has passed. Today we have multi-national giants who are much too large to be taken piecemeal by the nationalisation programmes of varying nations and ideologies. It can only be done by internationalisation. But how can we have internationalisation when we cannot even get agreement on the EEC? The effect of these different political aims and ideologies is to nullify corporate strategy. The only emphasis that can be placed upon these multi-national giants is co-operation at all levels. *Co-operation in planning for the future.* What are the requirements of the company and what are the requirements of the nation? How can these attitudes be merged? Professor Perimatter predicts that in the eighties over half the needs of the Western world will be provided for by a mere 200 companies. Who can visualise the internationalisation of these companies by the co-ordinated self-sacrificing Western nations, politically and economically?

Corporate strategy today is stuck for the most part with the short-range planning which can be accurately forecast by modern sophisticated techniques, whereas the long-range forecast for corporate strategy has hardly taken off the ground because of the lack of training and understanding of technological forecasting.

## Technological Forecasting*

Scientific crystal ball gazing is not a new concept. Eminent men have cast their eyes regularly on the future over the past several hundred years. At the present time we are within a period of great technological change and growth, with all its many consequential interactions upon the sociological environment of the industrial nations. During the past hundred years the technological changes that have taken place have far exceeded that of any technological change in the preceding 5,000 years. Since the beginning of the century the time it takes from the emergence of an idea to its crystallising into a useful technological form has been cut by over 60 per cent. For example, from the first historic manned plane flight until the

*Technological forecasting is the process of estimating as a function of time in the future, the amount of technical information which will exist in a well defined field of activity. (*The Prospect of Forecasting Technology*, Lawton M. Hartman.)

time when planes were flying at 400 miles per hour it took a technological life span of 30 years; to double that speed it took only 10 years and to reach a speed of 1,800 miles per hour it took a further life span of 5 years. From this point of view alone it should be apparent to management that we must pay considerable attention to any technological change which can effect our industries. What are the interactions and to what extent is industrial, demographic and sociological analysis geared, to be able to forecast future trends within our environment? Olaf Helmer of the Rand Corporation has estimated that there will be 25 million scientists and technologists in the world by the end of the century. If this is true, we can only presuppose that with the present number of scientists and technologists we are still only on the brink of technological change. Today more and more companies are looking towards technological forecasting as a management tool to assist them in formulating corporate strategy. The assumption that having a R. and D. department automatically presupposes that this discipline is catered for by the department is a thing of the past.

R. and D. are too busy and in many cases capitally unable to even meet the requirements of decisions and strategies already made. Although R. and D. is much concerned with any future technological change, they are in the main totally restricted to a 1–5 year horizon of product development, refinement and new applications, whereas technological forecasting is involved in the longer term future, the how, when and where aspects of the company's survival. Some companies have already successfully carried out technological forecasting programmes. Where these have been carried out, the company has either:

1. Had a dependency upon technical leadership.
2. Have been fortunate in having the right men whose technological forecasting fits into the organisational pattern.
3. That they have a technological forecasting programme which has been kept relatively simple.

Alternatively they have had:

1. A management who are sensitive to opportunities.
2. Obtained "Bootleg" research and the "Tinkering" of men close to products and operations.
3. Developed through market pressures and opportunities which have fallen their way.

Is there any advantage to be gained by any long-term technological forecasting? To many there would appear to be no need to do so. How often does one hear the words discount cash flow, capital budgets, etc., bandied about? We allege and presume that we plan, but we know that we do not. Our budgets are fictitious and we know it; we have never had any faith in any long-term planning. Fortunately management training is overcoming this tired mental attitude and the realisation of the needs for objective planning and the fact that this can be resolutely achieved is making way for a deeper understanding of corporate strategy. Technological forecasting is a single discipline within the field of corporate strategy which is becoming known simply as T.F. In order to carry out any form of T.F., however, we must fully understand what it is. When we consider the potential of technological forecasting for any company, we must have an analytical approach, e.g.

1. What is the purpose of T.F.?
2. How is it to be accomplished?
3. What are the capabilities and limitations of the methods to be used?
4. How can we organise for T.F. and who should be assigned to the task?
5. How can we be sure that we have any useful output?
6. Will it detract any time from useful production output?

All these questions have to be answered before we can approach the subject of T.F. To be useful, T.F. does not necessarily need to predict a precise form of technology. Its purpose is purely and simply to help to evaluate the probability and significance of various possible future developments, so that managers can make better decisions for their forecasts. For example, animal meat is anticipated to be an insignificant source of protein in the 1980s and the use of protein from the soya bean and new high intensive fish farming will take over in importance. A technology therefore is only utilised if it responds to a "need", otherwise it remains a capacity and never becomes a functioning reality. First we must have a creative feasible idea, then move on to its practical application and finally diffuse the technology throughout society. We must have a purpose in forecasting if we are to achieve the successful diffusion of a new technology through society. We must set a goal or objective, then identify the opportunities and also the threats to the goal or objective. By threats to a goal or objective we do not only mean the obvious threat from one's competitor, but also from the limitations of forecasts, by unpredictable interactions, unprecedented demands, major discoveries and/or inadequate data. When it is considered that a natural time unit in social life is one generation and an average regeneration occurs during the 20–25 year span it could be safely assumed that a normal regeneration cycle is one which technological forecasting could accept as a basic time frame, this time frame being broken down into either "singular" or five-year units. Time units, once set within a strategy, would set an organisational pattern for technological forecasting. Each technological forecast should be a prediction, with a stated level of confidence for the anticipated occurrence of a technological achievement, within a given time frame coupled with a specified level of support.

Organisational approaches to technological forecasting have been taken by the following forms:

1. Scientific advisors.
2. Wild men, i.e. talented imaginative and active individuals who stimulate new thought.
3. In company staffs.
   3.1. Staff planning or programme evaluation groups.
   3.2. Long-range planning groups.
   3.3. Opportunity-seeking groups.
   3.4. technical information centre and/or commercial intelligence units.
   3.5. Brain storming.

Technological forecasting can be distinguished from something called "an awareness of technological changes in the environment" inasmuch as it involves the systematic handling of information components, so that the combined data may disclose more than the separate parts.

The principle fundamental techniques of T.F. are as follows:

1. The Delphi method.
2. Morphological analysis.
3. Scenario writing.
4. Envelope forecasting.

## 1. The Delphi Method

The Delphi or Delphi Oracle method is the consensus of opinion from several experts and this is built up from a normal development of systematic probes into the future. The first probe is a prediction of the important events occurring in the area in question, from each expert in the group, in the form of brief statements. Subsequent probes are the clarification of the first probe by the investigator, followed by successive individual requestioning of each of the experts, combined with feedback supplied from the other experts via the investigator.

## 2. Morphological Analysis

Morphological analysis is a technical forecasting technique where all the energies are consolidated into a systematic approach, where all possible solutions are correlated, defined and analysed in relation to a particular problem.

## 3. Scenario Writing

By writing a scenario one provides a description of all possible futures and so as an aid to thinking it serves to call attention, sometimes dramatically and persuasively, to the larger range of possibilities that must be considered in the analysis of the future. The scenarios force the analyst to deal with details and dynamics that he might easily avoid treating if he restricted himself to abstract considerations.

## 4. Envelope Forecasting

The envelope curve represents the value of the technical attribute over time, independent of the devices used to achieve them. Each attribute for a given device has its value laid out upon a graph, and similar graphs are made for all individual devices in the same technical attribute, e.g. Travel. The subsequent curves plotted are then laid out upon a time grid.

These, of course, are only a few of the many technological forecasting techniques. In fact, Erich Jantsch claims that there are at least a hundred techniques, any of which could be used before coming to a particular decision on the type of forecast to be used for a particular case.

To be able to forecast we should also have:

1. Background: An organisational goal.
2. Present status: The current state of the technology, quantitatively described.
3. Forecast: The projection of the technology with its identifying parameters.
4. Product implications: The effect it will have on the company.
5. References and associated activities: It is this breadth of technological assessment which allows the forecasts to be beneficial to a diverse audience.

There are many pitfalls to technological forecasting which must be avoided if the techniques are to be successful.

These pitfalls are:

1. *Lack of Imagination.*
   1.1. Acceptance of trends rigidly continuing.
   1.2. Reliance upon limited information.
   1.3. Concentration on specific configurations of technology now in use.

2. *Mistakes in Methodology.*
   2.1. The use of inappropriate techniques.
   2.2. Inability to handle the uncertainty intrinsic in forecasting technology.
   2.3. Reliance on only one opinion.
   2.4. Over-reliance on extrapolative, quantitative procedures.

3. *Failure to Use any Formal Technique.*

Most companies today depend upon the information from customers, suppliers, competitors and other diverse sources to maintain an up-to-date assessment of the technological environment; intuitive and group assessments abound. In actual practice the techniques of T.F. are fraught with numerous obstacles and widespread managerial resistance. Managements must carefully consider their needs before initiating a firm forecasting programme. A management who fails to appreciate the changing face of its technological environment may miss many profitable opportunities, or will ignore major threats to their existence. Today's accelerating pace of product innovation and product obsolescence results from the technological explosion affecting almost all industries through changes in products, production processes and/or imput components (including also human resources).

Technological forecasting is no panacea, but it tells you where to look, it provides insights into the time relevancy of innovation. A significant area to be examined for T.F. is the area of environmental control; where is technology controlled by the environment? Management is working in an environment and looking into the future! Societies and nations are changing from the pre-industrial into the transitional; transitional into industrial; industrial into mass consumption and mass consumption into post-industrial.

Where any attempt to forecast a technology is to be either "Heuristic",* "Propaedeutic",* or "Paradigmatic",* with many cross-cultural comparisons, therefore, is management to integrate any form of T.F. with that of long-range planning? T.F. challenges the difficulty of interpreting and communicating the potential impact of an observed change. R. and D. projects must be objectively selected, there must be no staff line barrier and technological forecasting and planning staffs must include intelligent, creative individuals, in fact personnel who can interpret the potential impact of change. Technological forecasting is a discipline of management to project into the future and towards this end there is the development of first and second primary data.

*First Primary Data*

Growth is considered in terms of cumulative time or calendar year. Typically a plot of functional capability versus time displays the exponential character of growth (see Fig. 12).

*Second Primary Data*

The second set of primary data is concerned with value measurement trends in the techniques, which enables the functional capacity to be

*Heuristic – the art of serving, to find out.
  Propaedeutic – the subject of study which forms an introduction to an art or science.
  Paradigmatic – writing to serve as an example to others.

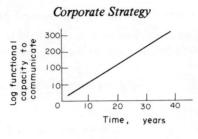

Fig. 12.

accomplished. This data considers specific techniques and this class of data follows a characteristic curve (see Fig. 13). Initially the technique experiences a slow growth, the potential is recognised, money and work is poured in, problems are resolved and the growth accelerates until finally limiting factors are encountered and the growth decelerates and the curve reaches its upper value or limits.

The curve of the second primary data is sometimes referred to as the 'S' curve or biological curve with limits determined by:

1. Laws of nature.
2. Competitive forces.

When forecasting for an uncertain technology, where time is the critical parameter, it is normal to place a confidence limit band to the initiated curve (Fig. 14).

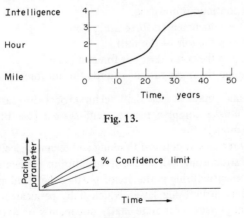

Fig. 13.

Fig. 14.

Where the value of the pacing parameter is the critical factor in the level of confidence, a spread of time is made for the given value (Fig. 15).

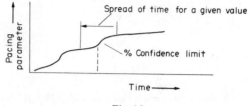

Fig. 15.

Technological forecasting follows a pattern and carries itself through a system of events, from a signal to a need, prediscovery to discovery, to creation, to substantiation through development to advanced engineering, to application and service engineering. All these are logical stepping stones of events from a conceived idea to creation; the technology of tomorrow is always with us today, if we heed the signals. Therefore, there is a need for an organisation for forecasting and we must, in organising, plan a system to suit the needs of the type of forecasting to be done. Therefore we should:

1. Plan the forecasting period.
2. Decide upon what we shall call it.
3. Decide upon the hierarchical nature of our forecasting.
4. Plan for continuous updating.
5. Formulate a documented forecasting plan.
6. Decide upon the role of the staff.
7. Decide upon the content of our forecast plans.
8. Decide upon the organisation required for our forecasting.

By this, one may expect that technological forecasting will be incorporated into a complex management system (see Fig. 16 for an organisational model).

Forecasting requires functional thinking and technological forecasting is a problem of structuring. All levels of structuring permeate all levels of effort; therefore structuring is the art of putting the mind and communication in order. Structuring levels which T.F. permeates are those of corporate management, corporate staff, documents or data centre, and divisions. Strategic areas are assessed, key objectives or areas are

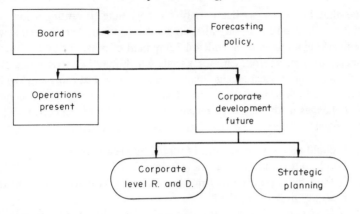

Fig. 16.

formulated, forecasting plans made, approved, reviewed, reassessed and reapproved; each plan should not exceed any period of one year without being reapproved. A minimum period of strategic technological planning should be seven years, as any shorter period of time cannot be either economic or of advantage to the company. Besides organising the company and setting a time scale, there is the necessity of monitoring the environment for activities which may effect any forecast.

*Monitoring*

1. Searching the environment for signals that may be forerunners of significant technological change.
2. Identifying the possible consequences (assuming that these signals are not false, and the trends they suggest do persist).
3. Choosing the parameters, policies, events and decisions that should be observed and followed to verify the true speed and direction of technology and the effects of employing it.
4. Presenting the data from the foregoing steps in a timely and appropriate manner for management's use in decisions about the organisation's reaction.

In the attainment of achieving a technological forecast one must realise that this is only a crossing point in time, e.g. one invents plastics only to realise that inventing plastics is only the beginning. Inventing a form of plastic is a key point. Upon enlarging its development one finds that we are embarking upon an open-ended development of possibilities.

Searching the environment for signals which may be forerunners of a significant technological change may be easy to say, but can we recognise them? If we look around us, we could see what may be signals for technological changes in the home. For example, we could take the cooker. What technological changes could effect the cooker as we know it?

1. It could take on a new form of supply, nuclear perhaps.
2. Food itself could take on a new form.
3. If food takes on a new form, would it be instant food, without the necessity of being cooked?
4. If food still had to be cooked, could a heat chamber be an integral part of the central heating system?

The same can almost be said for refrigerators. Are they going to remain more or less as they are, or will a cold room become an accepted part of the home? We take so many things for granted, that we do not see the signals for future development, signals which are being formulated by society and the environment. It is assumed that washeterias, washing machines, soap powder and laundries are going to remain with us forever, but in the future these may not be necessary. Clothing could take on a new form, which, coupled with the demands made by society for clean air and anti-pollution acts, could hasten the development of dirt-repellant clothing which never requires washing, or on the other hand clothing may be made of disposable material. The shortage of energy and the public acceptance of the coal face worker's job as being a dangerous and unnecessary one, in the light of the discovery of methods to contain energy within the pit without the necessity of mining, and the construction of costly tidal flow stations may become acceptable features. The signals are already there, but can we identify them and having identified them shall we have the courage to act upon them and intelligently forecast the technological parameters attached to the signal area? Should we be able to do this, we have to identify the correct parameter and strategy to follow. Management has to give permission to pursue the strategy, obtaining a technological success if possible, having made sure that we have correctly read the data in our forecast. Every signal which may be followed will not result in having a success. Therefore the T.F. function must be such as to show up the various parameters and be reassessed at each stage, shelving where necessary. The environment will take upon itself a force of change towards technological forecasting, as the sociological and economic demands within the environment will be a deciding factor for management, either to accept T.F. or become a part of the backwash, attempting an existence upon what remains.

Those who succeed will be managements who have a viable technological forecast based upon a true assessment of the realities of the changes which have taken place or are about to take place.

CHAPTER 15

EDUCATION AND THE ESTIMATOR

Jean Floud and A. H. Halsey, in the introduction to *Education, Economy and Society,* stated that "Education is a crucial type of investment for the exploitation of modern technology. This fact underlines recent educational development in all major industrial societies. Despite idiosyncracies of national history, political structure and social tradition, in every case the development of education bears the stamp of a dominant pattern imposed by the new and often conflicting pressures of technological and economic change."

It is a significant fact that such prominent sociologists find it crucial to have technological education in industry developed. Underlying this development must also be the education of the estimator, an education which many experts consider it is not at all possible to commence until the minimum age of 25. By this time it is considered that one has been able to prepare a sound basis of a theoretical and practical nature, able to assimilate and undergo a fuller training as an estimator in the manner which estimating should be regarded, as a true art or science. It is one of the misfortunes within industry today that no educational facilities exist purely for the benefit of the estimator and that there are no known examinations for the specific science of estimating, by which standards could be set and an estimator could be judged, for his ability and suitability in the estimating function. Today a small start has been made by the Association of Cost Engineers, which covers approximately 30 hours of lectures upon estimating (as an endorsement course to the engineering H.N.C.). In addition, some technical colleges do run an odd lecture or two upon estimating, within courses for other disciplines.

The inability of companies to obtain fully trained personnel in estimating is one of the sad reasons for the decline of financial solvency within industry. The estimating function, which should be the forte of fully trained personnel has become the dead-end job to which someone usually has to be appointed, or relegated. Most managements feel that the position of an estimator is a low priority, and have yet to realise that the company only survives because of the information supplied to management by the estimating department. Sales or marketing are unable to exist without the adjudication of work which can only be carried out by the functions of a resolute estimating department. The very nature of throwing personnel into the estimating function, usually upon low rates of pay, is one of the reasons why the general movement of estimators is from one company to another in search of a higher reward, or the threat of moving resulting in a new offer of just sufficient remuneration to make the estimator stay. There is then, because of this movement, no continuity in the standard of the estimating function.

Estimators for the Ministry of Technology have to have a university degree (any kind). The ministry believe that they can educate and train them in their estimating methods, but without a fuller experience of engineering it is understandable why their estimates have been so inaccurate. The Ministry of Defence works along similar lines, but they withdraw officers from their regiments to do a stint at the ministry, again rather a hit-or-miss affair. It is no wonder that public money is being continually wasted alongside that of industry; it is this lack of specialised training which leads to all the under- as well as grossly over-estimating of projects, which along with the monetary losses can be coupled the loss of effort by fully qualified development teams. Everyone expects development teams, scientists, tool room operatives, even the press operator to be trained for the job. Why not estimators? Industrialists, civil servants and politicians should realise that the time has come when money is all-important and that it is other people's money that they are spending. The field of estimating is becoming one of paramount importance and one will not be able in the future to get by simply upon a cost evaluation given by a commercial accountant. It is the estimator tackling the problem from the grass roots who will be able to give a correct assessment and evaluation of a problem. He alone is able by his expertise to know where to allow for work involved in a problem and where to obtain all the necessary data and information required, also when to obtain specialist information relating to a problem.

The estimator, although working as an individual, is basically always working as part of a team, with the expertise to evaluate the knowledge of the team into an accurate cash estimation of the work on hand. This form of work cannot be left to anyone: there must be some form of specialised training which must be implemented now, before we become desperately short of skilled and trained estimators. Up to the present time we have relied upon certain individuals who find themselves in the field of estimating, liking the work and so progressing with a satisfaction of their own in training themselves. Increasing this knowledge and experience in their own particular field and function, they become experts and are able to appreciate any difficulties regarding work placed before them for which they can estimate.

This knowledge and appreciation can only be as good as the training and experience in all the basic engineering functions which a particular estimator can obtain during his lifetime. The fact that a machine can operate at such or such a speed, that one man can erect equipment in so many days, therefore, four men can do it in a quarter of the time, or two machines in half the time, is irrelevant, as is the comparison of the costs of the R.101 with that of a lunar shuttle. Bare assumptions like these are simply not good enough and the job or function of the estimator is to be able to give as accurate an evaluation as time and information allow. It is impossible for an estimator to give the price of a hospital in 15 minutes or the cost of a *Queen Mary* in 10 minutes; there must be "think time". Very few companies realise that an important aspect of the estimating function is the "think time". A "with it" company has its think tanks, but to an estimator this idea is as old as the hills — without time, time in which to think, he is unable and incapable of doing an accurate job; yet how many companies will allow the estimator the time to evaluate drawings and problems correctly? Estimators need flexible, imaginative, enquiring minds, to be able to keep up-to-date, constantly learning, continually having to change ideas. It is time that the estimator came out of the backwoods and interchanged and discussed these ideas, without keeping in the background and being frightened to let anyone participate in his experiences. Because of this there is the need for a fully specialised course for the position of an estimator with a definite status attached to it, followed up by a system of refresher courses and the interchange of ideas and information. The major question is where can the education of the estimator start and what should it cover? Upon reflection of my own personal experience, a considerable amount of time has been wasted upon learning and acquiring knowledge which has been totally superfluous to the estimating function. It is always the same. One is wiser after the event, never before. Throughout the time I have been an estimator never can I recall any time where I was sent upon a course which was advantageous to me as an estimator. Neither did I receive any formal education or training in the field of estimating. Also I never became aware of any course of lectures which would have benefited me, until these last few years. Once I became interested in the field of estimating, I found it extremely interesting, as it is a continual challenge to one's intelligence, whether estimating for a small bracket, an industrial building complex or the bearings of the Taq Taq bridge in Iraq. The more I have become involved in estimating, the more I realise that it is not just a mere function within industry, but a science which could become more exacting and one which will require a definite programme of training before one can keep up with the modern managerial and quantitative techniques that are being undertaken in industry. It is a science in which one never stops learning. One is continually on the go searching for the best answer to the problem on hand. One does not just learn to estimate and stop there. The interaction of events, prices and materials are constantly in a state of flux and the changes which take place make it so that the estimator is constantly searching and probing for an answer. An important fact is to arrive at a standard and status for the estimator. There is at present too little understood about the actual functions of the estimator. Some have the functions of a cost engineer or project engineer, whilst others are termed estimators but are in reality only estimating clerks, carrying out a clerical estimating function of a repetitive nature, whilst others are true estimators. There has to be a clear understanding as to the actual function and duties of the estimator before he can be termed as one. When this fact is recognised, there would be a basis for education. Estimating is not merely the clerical function which many persons seem to think it to be, a job to be done by the first person to be thrown into the fray. With the evolution of time, the duties and functions of an estimator are becoming such that they will not be able to be carried out accurately unless by a fully qualified person. To this end there must be a complete change of attitude by management along with a complete understanding of the estimator's ability and the setting up of standards whereby the manager will know what type of person he is employing. If we are to have this change of attitude, it must also be necessary for the estimator to understand the functions of management. It will be the fundamental recognition of estimating as an art or science which will take time and which will be the main stumbling block in being able to start and sustain a complete educational programme. A small beginning has been made, and upon this nucleus must be built a future fuller educational programme. With the recognition of estimating as an art and the benefits that can be received by a company in employing a fully trained and qualified estimator, there will become because of this, recognition by management and a demand for these estimators. Once there is a demand, and the demand is seen to be there, technical colleges will put forward the necessary training facilities, even though these may be a little late. From a small beginning of enlightenment we must exchange information and ideas in order to make sure that the duties and functions of the estimator are more fully understood. These will vary in different industries and it is understandable why there is a great misunderstanding as to the proper duties and functions of the estimator. Primarily the duties and functions as understood by most estimators are as follows:

1. The full appreciation of an enquiry and all it entails.
2. From this appreciation, evaluate all the manufacturing involvement which it covers along with all the necessary labour content.
3. To be able to place a value upon all the requisite component parts.
4. Cover by a correct assessment, using one's own expertise, any eventual changes in the value of labour, materials or overheads.
5. To be able to give to the sales department an accurate price for the work or article, that was called for in the original enquiry, so that the sales department can place some reliance upon its accuracy.

When the estimator has been given a clear brief by the company for his estimating function, providing that he is trained and experienced in these duties, we will get away from the debacles and inaccuracies that have continually hounded estimating. We should be in the happy position of being able to evaluate, appreciate and analyse all situations in order to ensure that inaccuracies, etc., do not happen, learning, not criticising and passing it off purely as incompetence. Somewhere there has been a fundamental failing. This failing is the inability of the estimator to be trained or receive any full training in what should be his proper function. This fundamental failing must be stopped and prevented so that we can carry on into the future and survive. It is important that there is a correct period of training along clear and determined lines and thus train minds to be able to probe, search, absorb and analyse, in order to be able to ascertain and judge what is going to be the best method of manufacture and arrive at the keenest and most competitive price. Now that we have entered the Common Market it will be of paramount importance to this country for estimators to have this ability. Considerable thought should be given to the training of estimators now, and suitable candidates must be given the chance to begin early in their careers. Candidates should go through a full course of training and instruction which is clearly recognised by the educational authorities as well as by industry and all concerned. It should at least cover the following:

## Suggested Syllabus for a Course in Educating and Training the Estimator

1. An induction into estimating.
2. Drawing technology and appreciation.
3. Engineering and workshop practice.
4. Industrial organisation.
5. Accountancy and company law.
6. Objective training.

### 1. An Induction into Estimating

The induction should cover information about his own company and its estimating methods.

1.1. The size of the company, history and anticipated growth.
1.2. The company products, market and competition.
1.3. The company organisational structure; communications within the company and how one fits in.
1.4. The company's personnel and training facilities, future outlook and policy.
1.5. The union set-up and negotiating procedure within the company.

### 2. Drawing Technology and Appreciation

2.1. How to draw and translate drawings.
2.2. The work of the designer, draughtsman and tracer.
2.3. The use and application of drawings relating to specific associated industries and processes.

2.4. The use and application of drawings within his own industry, as well as any particular techniques.

2.5. Calculation of material from drawings.

## *3. Engineering and Workshop Practice*

3.1. The student estimator *must* have a thorough grounding in the use of, applications and operating of machines and equipment, also the techniques employed in the manipulation of materials. This must cover all the aspects of the type of work carried out within a company, with not only the theoretical use but also the practical.

3.2. He must also be able to recognise the suitability of materials used within the industry, the limitations of their use and any specific difficulties that can arise out of the use of particular materials.

3.3. The student estimator must have experience on the shop floor and in the workshop and be able to understand the functions and operations of each.

3.4. Students must also be given a period of time in all departments in order to understand fully the organisational function of the company and the contribution given on the various planes; in doing so he or she will be able to arrive at an appreciation of the functions and inherent usefulness of each department.

## *4. Industrial Organisation*

4.1. The student by now should have obtained an awareness of the functional organisation and should have already begun to understand the workings of the various departments concerned (as advocated in 3.4). He should also be aware of and know some of the functions of management and their responsibilities; in addition the responsibilities and functions of the board and management in relation to the employee, customer and shareholder.

4.2. He should be made aware of the techniques of investigation, forecasting, planning, co-ordination and control as effected by works and industrial action.

4.3. How employees are motivated to increase their skills and achieve better results as well as an understanding of the subject of human and industrial relations.

4.4. Methods of and channels of communication.
4.5. Welfare and social facilities and their effects.


## 5. *Accounting and Company Law*

It has been my personal experience that accountants like the assistance of estimators in presenting figures on engineering matters and they are only too willing to show the estimator how to present them; it would be of significant value if the estimator is trained in this field to do the following.

5.1. Estimators should be competent in preparing budgets for accounting purposes from given information and the preparation of data for cost evaluation. They should therefore understand the manner in which these should be prepared for accounting purposes.

5.2. They should also be aware of the implications of company law as it relates between buyer and seller (the law of contract), so that estimates are prepared in a manner in which there can be no doubt as to their meaning, particularly where the law of contract could apply.


## 6. *Objective Training*

In order for an estimator to be continually efficient, he must have facilities to enable him to keep up-to-date with any progress within industry which could have an effect upon his methods of estimating. In order to do this he must have the confidence of management, which can only be achieved by the estimator himself showing management his true value. Therefore management will want to assist him in any form whereby his career benefits by further training, whether by a day's convention, a short seminar, or longer periods which would also in turn be of benefit to the company.

6.1. Training in logic and perception should be catered for, as an immensely large proportion of the estimator's function is the ability to analyse and judge methods of work and performance. Towards this end it must be made possible for the estimator to receive the best training in the use of his mind logically to enable him to approach his work in a clear concise manner, evaluate and record clearly. Training in this field should also cover the statistical and quantitative techniques, as it is upon the ability to evaluate all the facts and record them clearly which forms the basis of a good estimate, from which one and all can fully appreciate what is taking place, what has to take place and the cost of doing this, providing information upon which a quotation can be made. Objective training, therefore, is twofold; it must be objective in giving the estimator a chance to keep up-to-date, it must also be the training which ensures that the mind is made and kept active and fertile. This is important. Estimating is not one of those functions which can be learnt and then left. The mind must have the ability to absorb knowledge, keep it up-to-date and throw it out when it has outlived its usefulness, absorbing fresh ideas and information to take its place; the brain therefore must be kept active and fertile.

The training of the estimator will be seen to be an extremely important step and one which must now be taken for the future. However, the number of estimators required per company is small, therefore parts of the training of the estimator which cannot be done by the company must be done by technical colleges. The training within the company of the apprentice must also be linked with existing basic engineering examinations, specialist training as an estimator being implanted upon them, until such times as an integral course for estimators can be started. The apprentice should also be made fully aware of what is expected of him. Too often apprentices shirk or bypass the unpleasant sides of their training, particularly where manual work is demanded of them. Training must be all round and must be fully completed by the apprentice, otherwise the balance is upset. The tendency of apprentices to get around the training officer and beguile their way into staying in the erstwhile comfort of an office rather than the manual experience of the works or shopfloor should at all costs be circumvented.

Training officers should not only be able to ensure that the apprentices complete their allocated periods in the various departments, but also discourage managers from asking to keep apprentices in their departments, as it becomes easy for departmental managers, who through pressure of work or other circumstances are short of staff, to try and retain the services of a good apprentice. We are unfortunately passing through an economic crisis whereby apprentice schemes are being curtailed, but is sufficient thought being given to the short-sightedness of this type of policy? John Vaisey and Michael Debeavais, writing about the *Economic Aspects of Education Development*, formed the following conclusions, which although are based upon the sixties are still relevant today. "We are driven to the conclusion that in order to meet the economic and social needs of the '1960s' Europe will have to give its people what, until recently, would have been thought of as a luxury education. This implies that education for many people will become longer, in order to increase their expertise and that it should become broader, in order to increase the possibility of contact with others and to guarantee them some freedom of movement as their specialisms become redundant or obsolescent. It should give everybody a general basis of culture, essential for economic growth and human dignity alike.

"How to achieve this without overloading the curriculum or making the formal process of education too long seems to be the central educational problem for the gifted. For the less gifted, the problem is to raise their educational status without either imposing intolerable intellectual burdens, or creating a gulf that does not correspond to their place in the spectrum of ability and that replaces the former economic and social division of society by one based on real or supposed talent. This poses complicated educational problems: economically, the questions are at least as complex.

"In the first place education is part of the expenditure of abundance, which is increasingly child centred and family based. In the second place, the economy requires an adequately educated labour force for its production capacity. In the third place, the bulk of the expenditure falls on to the government account and consequently raises problems about the tax burden. Fourth, there is the need to find skilled manpower to provide the teaching force. Fifth, without economic growth, a better education cannot be provided. The relationship between the economy and education is reciprocal."

Such conclusions can, therefore, only support the fact that industry must make a considerable contribution in order to be able to have their own experts trained to the standards to which they require them to be. They must also, in turn, state their understanding of each job analysis and the training specifications for the job described. It is becoming increasingly more obvious that in a technological age the training of manpower for the more technical jobs will have to be started at an earlier age. The present manner in which children are, in effect, graded 'superior', 'mediocre', or 'poor' in intelligence and attainments at the early age of 11, educated for a few more years and turned over to industry, either as manual or non-manual workers, whatever their education has dictated, will have to be drastically overhauled.

In *Education, Social Mobility and the Labour Market,* Jean Floud and A. H. Halsey state: "Some 20–25 per cent of the age group of 11 year olds is selected for education at grammar schools (though for particular districts the proportion may vary as little as 10 per cent to as much as 50 per cent and the provision bears no necessary relation to the needs of the local occupational structure). It is sometimes suggested that this minority, represents an intellectual élite – a broad based aristocracy of brains. It is true that it probably contains the best brains of the social classes who do not send their children to independent schools, but it also represents virtually the entire reserve of potentially qualified manpower. Less than 10 per cent of the age group is admitted at 11 to technical schools and a still smaller fraction at 13. The remaining two-thirds of the age group continue their education in the secondary modern schools, leaving for the most part at 16, to form the bulk of the supply of manual workers, very few of whom are destined to climb subsequently into posts of more than minor responsibility in industry or commerce." It follows from these assumptions that the system must be changed, as it is only from these sources that we are to obtain the extra technologists who will be required in the future. Any children who have the signs or appearances of being able to be trained on technological subjects must be given the opportunity at the earliest stage to receive the training necessary in technical schools. The nineteenth century saw the need to educate the illiterate masses in order to obtain the source of skilled labour that the Industrial Revolution was demanding. The twentieth century therefore demands more for the technological age.

The time has also come for the position of the estimator to be clearly defined, for training courses to be established and with subsequent examinations to uphold a status that is recognised by all. Unfortunately this does not only apply to estimators but to all vocations. The shortages will only be overcome when educational establishments are reorganised upon a technological field, letting those who wish to leave school at the age of 14 to become manual workers to do so. The subsequent savings in education can then be channelled into training upon a vocational basis for the requirements of a technological age from the ages of at least 12 to 18. By doing so we should be able to make far greater use of our limited resources.

# GLOSSARY

**Approximate Estimating.** The process of determining what the cost of a project should be. It occurs before the preparation of the cost plan and should be based on the client's fundamental requirements (James Nesbit).

**Bid Estimate.** A forecast of capital expenditure with an accuracy of up to ± 10%. Sometimes called: authorisation, project cost control, definitive estimate, class 2.

**Capital Cost Estimate.** A forecast which is prepared to calculate the profitability of a project.

**Contract Estimate.** A forecast prepared for a project based upon a final design.

**Contractor's Estimate.** Submitted by a supplier to a purchaser quoting rates or terms and conditions under which the supplier is prepared to execute or negotiate a formal contract.

**Control Estimate.** A forecast of capital cost prepared and used during the course of a project to indicate and check the probable total of final expenditure.

**Definitive Estimate.** This is a comprehensive forecast of capital expenditure prepared from well-defined engineering data. The engineering data includes as a minimum, fairly complete plot plans and elevations, piping and instrument diagrams, one-line electrical diagrams, equipment data sheets and quotations, structural sketches, soil data and sketches of major foundations, building sketches and a complete set of specifications. This category of estimate covers all types from the minimum described above to the detailed definitive type which would be made from "Approved for construction" drawings and specifications. It is expected that an estimate of this type would be accurate within ± 5%. Sometimes called: detailed, firm, project, cost control, or class 1 estimate.

**Detailed Estimate.** See "Firm price estimate" and "Definitive estimate".

**Estimate Authorisation.** Approval to expend within budgeted limits.

**Estimate Evaluation.** The adjudication of alternative offers to select the economic optimum.

**Estimating Rates.** The cost of unit quantities of material, labour, services or other factors, determined by experience, practice or policy, for the purpose of forecasting project costs.

**Factored Estimate.** A forecast of project costs made by applying multiplying factors to the major items in order to evaluate the minor components.

A capital cost forecast of a project when information is limited to a pictorial flow sheet and a geographical location compiled by the use of percentages for various ancillaries, such as electrical work, instruments, pipework, etc., applied to a basic figure comprising the total forecasted cost of a fundamental process equipment.

**Final Estimate.** A detailed forecast of a project cost, of sufficient accuracy to enable a contract to be prepared, or suitable upon which cost control could be based.

**Firm Price Estimate.** A forecast of fixed investment such as would be made by a lump sum contractor for bid from final drawings, specifications and site surveys. This forecast would be subject to fluctuation clauses from the base date, but could form a basis for the supplier to submit his firm price tender.

**Guesstimate.** A quick, order-of-magnitude estimate used in deciding whether or not to spend further time on analysis of a new idea.

**Horseback Estimate.** Forecast of an approximate project cost, which contains a high probability of error.

**Hypothetical Cost.** An estimate or expense based upon a theoretical assumption.

**Interest Rate of Return.** The calculated cash flow at which the company's average outstanding investment is repaid by proceeds of the project. Also, the evaluation technique which judges the profitability of a proposed investment by this calculation. Also called investor's method, profitability index, or internal rate of return.

**Order-of-magnitude Estimate.** An approximate forecast of fixed investment obtained without flow sheet or detailed equipment analysis by applying factors, ratios and escalation to published data or cost of a previous installation considered similar or by "turnover ratio" concepts. The least accurate of the estimate types, on which no confidence limits can reasonably be applied.

**Preliminary Estimate.** A forecast of expenditure with accuracy of ± 30% for rough calculation of a project cost and profitability, not sufficiently accurate for final approval. Sometimes called "study", "predesign", "evaluation", "order of cost", or "class 3 estimate".

**Present Value.** (a) A sum which if invested now with compound interest will provide a return of capital and interest in a number of years.

$$= \frac{(I + r)^n - I}{r\,(I + r)^n}$$

where $r$ = interest rate per cent divided by 100, and
$n$ = number of years.

(b) The discounted worth of a series of cash flows at any arbitrary point of time. Also, the system of comparing proposed investments which involves discounting at a known interest rate (representing a cost or a minimum acceptable rate of return) in order to choose the alternative having the highest present value per unit of investment.

This technique eliminates the occasional difficulty with Productivity index of multiple solutions, but has the troublesome problem of choosing or calculating a "cost" of "capital" or minimum rate of return.

**Profitability Index (PI).** The rate of compound interest at which the company's outstanding investment is repaid by proceeds from the project. All proceeds from the project, beyond that required for interest, are credited, by the method of solution, toward repayment of interest by this calculation. Also called discounted cash flow, interest rate of return, investors method, internal rate of return. Although frequently requiring more time to calculate than any other yardsticks, it reflects in a single number both the cash and the time values of all money involved in a project. In some very special cases, such as multiple changes of sign in cumulative cash position, false and multiple solutions can be obtained by this technique.

**Project Control Estimate.** A forecast of fixed investment prepared from completely specified equipment lists, finished engineering flow sheets, final plot plans and general arrangements, together with reasonably complete site and auxiliary facilities information. Such estimates should have confidence limits of ± 5% and are in sufficient detail to be the basis for sound job cost control.

**Rate of Return on Investment.** The efficiency ratio relating profit or cash flow incomes to capital employed.

**Return on Average Investment.** The ratio of annual profits to the mean book value of fixed capital, with or without working capital. This method has some advantages over the return-on-original investment method, depreciation is always considered and terminal recoveries are accounted for. However, this method does not account for the timing of cash flows and yields answers that are considerably higher than those obtained by the return-on-original investment and profitability index methods. Results may be deceiving when compared, say, against the company's cost of capital.

**Sanction Estimate.** A forecast of the capital cost of a project, which is sufficiently accurate to justify authorisation of capital expenditure or investment.

**Study Estimate.** A forecast of fixed investment or capital cost prepared from limited information of operating conditions and therefore is not regarded as an accurate or firm estimate. Sometimes described as "feasibility estimate".

The definitions given in this glossary are those standardised by the Association of Cost Engineers, and have been extracted from their publication *Cost Engineering Terminology.*

*Cutting off Chart*

| Length, mm | Width, mm | | | | | | | |
|---|---|---|---|---|---|---|---|---|
| | 0.30 | 30—45 | 45—60 | 60—75 | 75—90 | 90—105 | 105—120 | 120—135 |
| 400—600 | 1,500 | 775 | 750 | 700 | 650 | 615 | 610 | 560 |
| 600—800 | 775 | 770 | 650 | 630 | 610 | 590 | 570 | 550 |
| 800—1000 | 650 | 620 | 575 | 550 | 525 | 500 | 480 | 450 |
| 1000—1400 | 625 | 600 | 550 | 500 | 480 | 450 | 440 | 430 |
| 1400—1600 | 600 | 550 | 500 | 480 | 450 | 430 | 400 | 375 |
| 1600—2000 | 550 | 500 | 450 | 430 | 420 | 380 | 360 | 350 |
| 2000—2400 | 500 | 450 | 425 | 400 | 380 | 350 | 330 | 325 |
| 2400—2700 | 450 | 400 | 380 | 360 | 350 | 330 | 300 | 280 |
| 2700—3000 | 400 | 380 | 360 | 350 | 320 | 280 | 275 | 260 |
| 3000—3300 | 350 | 320 | 300 | 290 | 280 | 260 | 250 | 240 |
| 3300—3600 | 345 | 300 | 290 | 280 | 275 | 250 | 240 | 220 |
| 3600—3900 | 340 | 300 | 280 | 275 | 260 | 240 | 220 | 200 |
| 3900—4200 | 325 | 280 | 275 | 260 | 250 | 220 | 200 | 180 |
| 4200—4500 | 315 | 275 | 260 | 250 | 240 | 200 | 180 | 160 |
| 4500—4800 | 300 | 260 | 250 | 240 | 220 | 180 | 160 | 140 |
| 4800—5100 | 280 | 250 | 240 | 200 | 190 | 160 | 140 | 120 |
| 5100—5400 | 260 | 240 | 200 | 185 | 175 | 140 | 120 | 110 |
| 5400—5700 | 240 | 200 | 180 | 175 | 165 | 120 | 110 | 100 |
| 5700—6000 | 220 | 190 | 175 | 160 | 140 | 110 | 100 | 95 |

# APPENDIX 2

## German Technical Translations

Abgew Breite: developed width.
Anbiegen: verb "to bend".
Alle radien in blechebene gemessen: all
   radii measured in plane of sheet.
Ansicht in richtung 'A': view in direction
   'A'.
Ansicht: view.
Aus fahrsengmitte: from C/L of vehicle.
Blechstreifen: sheet metal strip.
Bis fahrsengmitte: to C/L of vehicle.
Bogan: arch.
Bis zur basis: to the base.
Dachbogenblech: roof arch sheet.
Dick: thick.
Fenster: window.
Fensterbogen: window arch.
Fahrtrichtung:direction of travel.
Hinterer anschweibrahmen: rear
   welding frame.
Horizontalbogen: horizontal arch.
Höhe ü basis: height above base.
Links gezachnet: left drawn.
Mitt fahrzeug: C/L of vehicle.
Oben: above.
Rechts spiegelbild: right mirrored.
Spant: rib.

Seitenansicht: side view.
Längsträger rechts: long beam light.
Punkt: point.
Senklöcher: pierced countersunk hole.
Überstand oben und unten an
   eckfensterblende anbiegen (siehe
   ansicht "Z"): bend projection
   above and below onto the
   window corner shield or stop (see
   view "Z").
Umrib und schmiegen nach: Outline and
   curvature to.
Und blechstreifen: and sheet metal strip.
Unten: below.
Beim montisen der klamme darf die
   sierlviste nicht deformiert
   werden: When mounting the
   brackets the ornamental strip
   should not be deformed.
Dieses ende nach unten biegt
   umherausfallen der klammern zu
   Verhinderm: This end bends
   downwards so as to prevent the
   bracket from falling out.
Kanten: Edges.
Beschnitten: Sheared.

## Swedish Technical Translations

Alt: or.
Alumin: aluminium.
Alternativt utför ands: alternative to
   drawing.
Antal/bil: number off.
Axel vindruteforkare: shaft of
   windscreen wiper.
Bakskärm: rear window.
Bland list med clips (utan clips): plain
   moulding with clips (without
   clips).

Deformeras: deformed.
Dekorationslist: decorative strip.
Deforationslist kplsm-def bakskärm:
   decorative strip for rear window.
Bakdörr: rear door.
Efter: after.
Eloyerad: anodised.
Enl godk prov: in accordance with
   approved specifications.
Fastsättning: fixing.

Forutsatter att, vindruteforkaray larna
    förlanges 30mm: provided
    windscreen wiper axles increased
    30 mm.
Grupp nr: group number.
Gummitätning: rubber sealing.
Högglanspolorad: highly polished.
Ingår på smst: associated drawing.
Justeras: adjusted.
Klistras: adhered.
Korriktning: direction of travel.
Larosscriplåt: car body sheet.
Längd: length.
Listen: strip.
Lufskärmen: screen.
Mittplan: centre.
Monterad: mounted.

Nedvikes I dennaände för att hindra
    klammern ramla ur: this end
    bends downwards so as to prevent
    the bracket from falling out.
Och: and.
Om så erfordras: if required.
På: on.
Plättjockler: sheet thickness.
Platkanten: panel edge.
Sektion: section.
Skala: scale.
Skarp kant brytes: edge removed from
    sharp corner.
Under: below.
Visande: showing.
Vriden: turn.
Ytbehandling: surface treatment.

# A P P E N D I X   3

## NOMOGRAPH – FABRICATION (LABOUR) COSTS

Tube:  2" n.b. M.S. to B.S. 1387 Heavy Grade

| DESCRIPTION | WELD | TEE |
|---|---|---|
| Provide 2" n.b. flanged branch | £4.75 | £7.85 |
| Provide 1½" n.b. flanged branch | £4.01 | £7.30 |
| Provide 1¼" n.b. flanged branch | £3.40 | £6.85 |

| ELEMENT | SITE | TOTAL |
|---|---|---|
| Provide B.S.T. "E" flanged end | – | £1.25 |
| Flanged joint in lieu of shop B.W. | £1.15 | £1.35 |
| Flanged joint in lieu of site B.W. | £0.80 | £1.10 |
| "Mid-Line" flanged connection | £1.15 | £3.85 |
| Single valve assembly, "Mid-line" | £3.90 | £6.85 |
| Double valve assembly, "Mid-line" | £6.65 | £9.60 |
| Single valve assembly. "Line-end" | £2.75 | £2.75 |
| 2"/1½" Conc. reducer at "Line-end" | £1.50 | £2.55 |

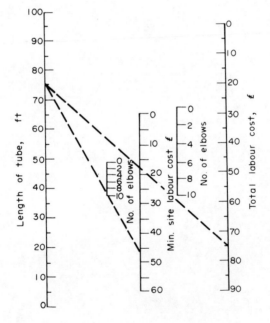

Example:

Tube length 75ft
incorporating 10–
90° welding elbows.
Total labour cost £75
Min. site labour cost £47

160

# APPENDIX 4

## NOMOGRAPH – MATERIAL COSTS

Tube: 2" n.b. M.S. to B.S. 1387 Heavy Grade

| DESCRIPTION ELEMENT | WELD | TEE |
|---|---|---|
| Provide 2" n.b. flanged branch | £0.35 | £1.65 |
| Provide 1½" n.b. flanged branch | £0.30 | £2.00 |
| Provide 1" n.b. flanged branch | £0.25 | £1.95 |
| Provide B.S.T. 'E' flanged end | | £0.35 |
| Flanged joint in lieu of shop B.W. | | £0.70 |
| Flanged joint in lieu of site B.W. | | £0.70 |
| "Mid-line" flanged connection | | £0.70 |
| Single valve assembly. "Mid-line" | | £0.70 |
| Double valve assembly. "Mid-line" | | £0.70 |
| Single valve assembly. "Line-end" | | – |
| 2"/1½" Conc. reducer at "Line-end" | | £0.90 |

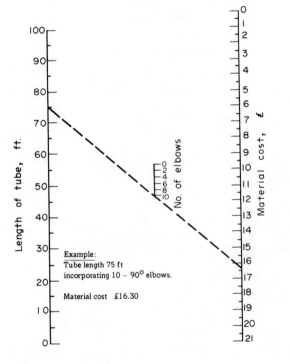

Example:
Tube length 75 ft
incorporating 10 – 90° elbows.

Material cost £16.30

161

# APPENDIX 5

## Basic Flow Patterns

'I' Flow

'U' Flow

'L' Flow

's' Flows

'O' Flow

# APPENDIX 6

## Inside Cylinder of a Steam Locomotive

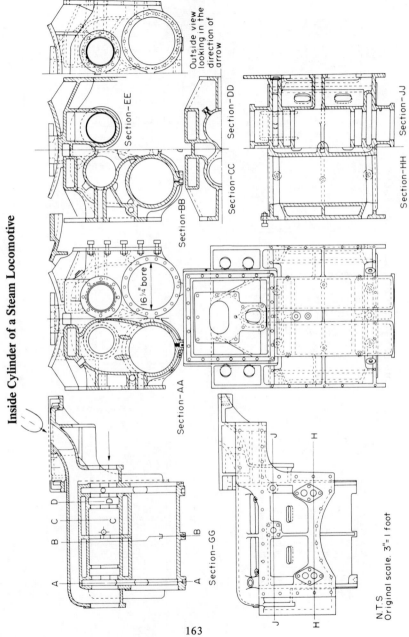

Outside view looking in the direction of arrow

Section-EE

Section-DD

Section-CC

Section-BB

Section-JJ

Section-HH

16¼" bore

Inside cylinders 16¼ dia x 28" stroke

Section-AA

Section-GG

N.T.S
Original scale. 3" = 1 foot

# APPENDIX 7

## Outside Cylinder of a Steam Locomotive

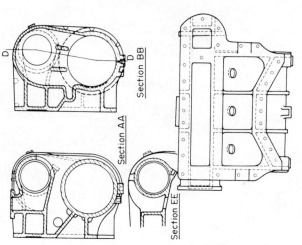

Section AA

Section BB

Section EE

Section CC

Outside cylinder 16 ¼" dia x 28" stroke

16 ¼" dia

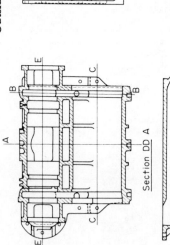

Section DD

N.T.S.

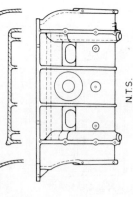

# APPENDIX 8

## Project Flow Sheet

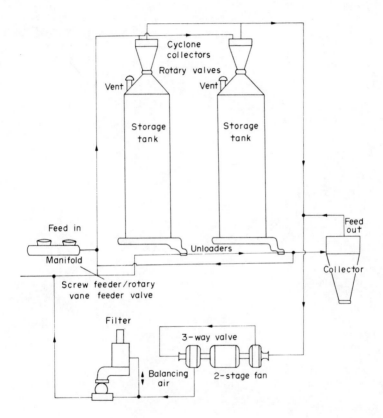

# INDEX